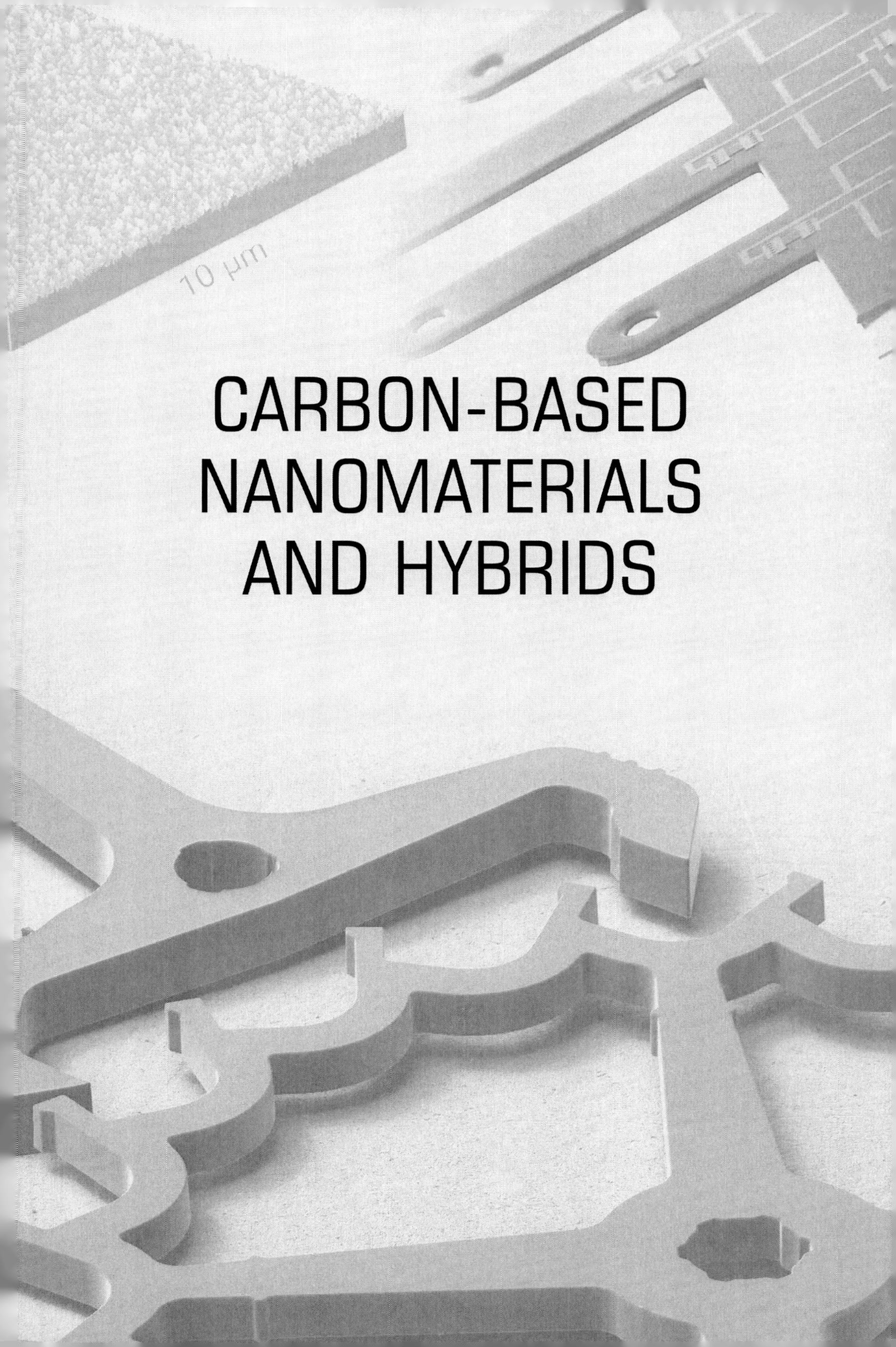

CARBON-BASED NANOMATERIALS AND HYBRIDS

CARBON-BASED NANOMATERIALS AND HYBRIDS

Synthesis, Properties, and Commercial Applications

edited by

Hans-Jörg Fecht | Kai Brühne | Peter Gluche

Published by

Pan Stanford Publishing Pte. Ltd.
Penthouse Level, Suntec Tower 3
8 Temasek Boulevard
Singapore 038988

Email: editorial@panstanford.com
Web: www.panstanford.com

British Library Cataloguing-in-Publication Data
A catalogue record for this book is available from the British Library.

Carbon-based Nanomaterials and Hybrids: Synthesis, Properties, and Commercial Applications

Cover image: The atomic force microscope image (AFM) demonstrates the ultra-smooth surface of a nanocrystalline diamond layer with a root mean square roughness of 16 nm (top left/courtesy Matthias Wiora, Ulm University).
Free standing piezoresistive (*n*-type) diamond microcantilevers with a thickness of typically 10 μm and a length between 500 μm and1000 μm (top right/courtesy Neda Wiora, Ulm University).
Diamond-based escapement wheel and anchor—the heart of a lubrication-free mechanical watch (bottom/courtesy Diamaze Microtechnology SA, La Chaux-de-Fonds, CH).

ISBN 978-981-4316-85-9 (Hardcover)
ISBN 978-981-4411-41-7 (eBook)

Printed in the USA

Contents

Preface

Carbon-based materials date centuries back in their synthesis and usage and comprise a whole realm of different crystallographic structures, chemical bonds and geometries, such as natural and synthetic diamond, different variations of graphite, carbon fibers, and their composites. Over the past few years however, the controlled reduction of sample size into the range of a few nanometers at least in one dimension has received growing interest and a renaissance of the field.

Diamond is renowned as a material with superlative physical qualities, most of which originate from the strong covalent bonding between its atoms. Diamond has the highest hardness and highest thermal conductivity of any bulk material and those properties determine the major industrial applications of diamond in cutting and polishing tools, windows, heat spreaders, and the scientific applications in diamond knives, diamond anvil cells, and as an optical detector material. Although diamond is thermodynamically less stable than graphite, the conversion rate from diamond to graphite is negligible at standard conditions.

Graphite generally can be considered as a well-ordered kind of coal and represents an electrical conductor, a semimetal that is mechanically rather soft due to its weak Van der Waals interlayer bonds and thus forms a two-dimensional structure. The basic unit of graphite is one layer of carbon, which is called graphene.

Furthermore, carbon nanotube or generally carbon fiber is a material consisting of fibers typically 5–10 μm in diameter. To produce carbon fiber, the carbon atoms are bonded together in crystals that are more or less aligned parallel to the long axis of the fiber. This alignment gives a high strength-to-volume ratio with carbon fibers exhibiting furthermore high stiffness, high tensile strength, low weight, high chemical resistance, and low thermal expansion. Carbon fibers are usually combined with other materials to form a composite. When combined with a plastic resin, it forms a carbon fiber–reinforced polymer that has a very high strength-to-weight ratio, is lightweight, and is extremely rigid although somewhat brittle with an abundance of applications in aerospace, automotive, and civil engineering; motorsports; and others.

Considering more recent developments, the miniaturization of material dimensions, components, and structures nowadays is reaching dimensions of a few nanometers—a development which generally is termed nanotechnology. In general, most materials properties are changed dramatically when reaching nanometer sizes and thus nanoscaled materials can be engineered through the controlled and size-selective synthesis of nanoscale building block with tunable and improved physical and chemical properties.

This trend over the last decades has been taken up here and represents the main focus of the present book applied to C-based materials. Nano-sized C-based materials include several modifications and geometries, such as nanocrystalline diamond, amorphous diamond-like carbon (DLC), C-based aerogels, and carbon nanotubes (CNTs), while some other new developments including fullerenes and graphene are still in their infancy. The book compiles and details cutting-edge research, and several applications are described within the fields of energy, microelectronics, biomedicine, and beyond. Furthermore, a perspective is given, including a diversity of industrial applications and market opportunities for C-based nanoscale materials and devices in the future.

With eight chapters contributed by world-class scientists and engineers, this book covers most recent developments in the science and technology of C-based nanomaterials for a number of industrial applications. It addresses both academia and industry research and engineering in this fast-developing field.

Hans-Jörg Fecht
Kai Brühne
Peter Gluche
Spring 2014

Acknowledgment

The financial and intellectual support of BMBF-VDI/VDE IT (C-HYBRID 16SV5320K and VIP-DiM 16SV6053) as well as Ulysse Nardin, Le Locle, Switzerland, and Audemars Piguet, Le Brassus, Switzerland, are gratefully acknowledged.

We would also like to thank the publishers, in particular, Stanford Chong and Shivani Sharma; and a number of graduate students working in the group, including M. Mertens, M. Mohr, N. Wiora, and D. Zhu for their excellent contributions; and Helga Faisst and Carolyn Kotlowski for their expert technical support.

Chapter 1

C-Based Materials on a Nanoscale: Synthesis, Properties, Applications, and Economical Aspects

Hans-Jörg Fecht and Kai Brühne
Institute of Micro and Nanomaterials, Ulm University, Albert-Einstein-Allee 47, 89081 Ulm, Germany
hans.fecht@uni-ulm.de

Nanotechnology is an emerging/enabling technology and of utmost importance. Two main approaches are used in nanotechnology. In the **bottom-up** approach, materials and devices are built from molecular components that assemble themselves chemically by principles of molecular recognition. In the **top-down** approach, nano-objects are constructed from larger entities with or without atomic-level control. The impetus for nanotechnology comes from a renewed interest in interface and colloid science, coupled with a new generation of analytical tools such as the atomic force microscope and the scanning tunneling microscope. Combined with refined processes such as electron beam lithography and molecular beam epitaxy, these instruments allow the deliberate manipulation of nanostructures and lead to the observation of novel phenomena.

Carbon-based Nanomaterials and Hybrids: Synthesis, Properties, and Commercial Applications
Edited by Hans-Jörg Fecht, Kai Brühne, and Peter Gluche

ISBN 978-981-4316-85-9 (Hardcover), 978-981-4411-41-7 (eBook)
www.panstanford.com

This newly emerging field includes the use of nanotechnology effects to achieve better device performance or to create completely new devices (bottom up). The trend to a continuous miniaturization and the corresponding increase in the density of integration is a challenge to the processes and materials in use (top down). Therefore, this exciting area at the interface between the micro- and the nanoworld is gaining more and more interest from the fundamental point of view as well as for industrial applications. One of the key scientific and commercial problems to be solved is the development of novel functional structures of superior performance by controlling the atomic or molecular structure on a scale between 1 nm and 100 nm.

As such, nanotechnology comprises.

- all products with a controlled geometry size of at least one functional component below 100 nm in one or more dimensions that makes physical, chemical, or biological effects available, which cannot be achieved above the critical dimension(s) (~100 nm) without loss of performance; and
- equipment for analytical or manipulatory purposes that allows controlled fabrication, movement, or measurement resolution with a precision below 100 nm.

Most fundamental physical properties change if the geometry size in at least one dimension is reduced to a critical value well below 100 nm, depending on the material itself. For example, this allows tuning of the physical properties of a macroscopic material if the material consists of nanoscale building blocks with controlled size and composition. Every property has a critical-length scale, and if a nanoscale building block is made smaller than the critical-length scale, the fundamental physics of that property changes. By altering the sizes of these building blocks and controlling their internal and surface chemistry, their atomic structure, and their assembly, it is possible to engineer properties and functionalities of materials and devices in completely new ways.

An overview of the effects and applications of the reduced dimensionality of nanomaterials is listed in Table 1.1. Furthermore, the addition of nanoparticles to an otherwise homogenous material can lead to a change in the macroscopic material behavior. Most material properties may be changed and engineered dramatically through the controlled size-selective synthesis and assembly of nanoscale building blocks.

Table 1.1 Effects and applications of the reduced dimensionality of nanomaterials

Nanoscale effects	Applications
High surface-to-volume ratio and enhanced reactivity	Catalysis, solar cells, batteries, gas sensors
Enhanced electrical properties	Conductivity of materials
Increased hardness from small crystal grain size	Hard coatings, thin protection layers
Narrower electronic band gap	Opto-electronics
Lower melting and sintering temperatures	Processing of materials, low sintering materials
Improved transport kinetics	Batteries, hydrogen storage
Improved reliability	Nanoparticle-encapsulated electronic components
Increased wear resistance	Hard coatings, tools
Higher resistivity with smaller grain size	Electronics, passive components, sensors

Systems with integrated nanometer-scale structures and functions present a multidisciplinary challenge. The performance of such microsystems also depends on an understanding of the properties on both nano- and microscales. Recently, the Review Committee of the National Nanotechnology Initiative in the United States recommended, "Revolutionary change will come from integrating molecular and nanoscale components into high order structures. To achieve improvements over today's systems, chemical and biologically assembled machines must combine the best features of the top-down and bottom-up approaches."

Chapter 2

Synthesis of Nanodiamond

Matthias Wiora, Kai Brühne, and Hans-Jörg Fecht
Institute of Micro and Nanomaterials, Ulm University, Albert-Einstein-Allee 47, 89081 Ulm, Germany
kai.bruehne@uni-ulm.de

The synthesis of diamond has been a dream of mankind since the 19th century when it was discovered that diamond is an allotrope of carbon. It was from this time that scientists eagerly tried to find an approach for conversion of the stable form graphite into the metastable form diamond. The first successful approach in 1953 was based on duplicating nature's method using conditions that are similar to a great depth within earth, typically pressures of several gigapascals and temperatures in excess of 1,200°C. The so-called high-pressure high-temperature (HPHT) method is well employed in today's synthetic diamond industry, producing hundreds of tons each year. Today, even a commercial production of high-quality gemlike stones of more than 3 carats is realized (e.g., Scio Diamond Technology Corporation or Brilliant Earth). However, due to its restriction in shape and size and infeasibility for integration into common microfabrication processes, the use of HPHT is mainly

Carbon-based Nanomaterials and Hybrids: Synthesis, Properties, and Commercial Applications
Edited by Hans-Jörg Fecht, Kai Brühne, and Peter Gluche

ISBN 978-981-4316-85-9 (Hardcover), 978-981-4411-41-7 (eBook)
www.panstanford.com

found in mechanical abrasion applications such as cutting, grinding, and polishing.

A much more versatile method for synthesizing diamond using a low-pressure chemical vapor deposition (CVD) technique was developed by different groups between 1962 and 1970 [1–3]. From that time on, this technique made it feasible to coat a variety of materials with diamond. In contrast to the HPHT method, yielding only single crystalline diamond stones, typical CVD diamonds consist of numerous small crystallites forming a continuous film. These polycrystalline diamond films can reach bulk properties similar to those of natural diamond.

This chapter gives an overview of the diamond deposition process, including the substrate pretreatment to enhance diamond formation, the basic principles of diamond growth, and the two most used techniques. A survey of the most important growth parameters and their influence on diamond's microstructure closes the chapter.

2.1 Substrate Pretreatment

The growth of diamond on nondiamond materials—irrespective of the method and crystallographic structure—implies an initial substrate pretreatment to promote nucleation. A combination of the high surface energy, low precursor sticking probability, and strong chemical competition between the carbon phases hampers primary diamond nucleation. Growth of diamond nuclei starts when individual carbon atoms on the substrate's surface begin to form sp^3 tetrahedral lattices. Since nondiamond substrates initially do not promote such a template for the tetrahedral structure, carbon atoms reaching the surface are immediately etched by the highly reactive atomic hydrogen. Consequently, diamond formation is not supposed to take place on ideal surfaces. In reality, even so, a low nucleation density of 10^3–10^5 cm^{-2} is detected. This can be mainly associated with the fact that surface defects such as steps, grooves, dislocation spirals, or grain boundaries, as well as dangling bonds and carbon-rich areas, are able to stabilize carbon sp^3 tetrahedral formation.

Even though spontaneous surface nucleation occurs on nonseeded substrates, the film evolution in terms of film uniformity, film adhesion, and roughness is considerably controlled by the initial nucleation density. Additionally, to ensure a minimum thickness at film coalescence, a high nucleation density is essential. The consequence of an improper nucleation density on the diamond film formation is shown in Fig. 2.1. An untreated blank silicon wafer is used as the substrate, and the film formation is analyzed by scanning electron microscopy (SEM). A significant, long growth period of eight hours is necessary to obtain noticeable diamond nucleation. In the figure, the spherical clusters consist of nanometer-sized nanocrystalline diamond (NCD) crystals, having fairly large diameters of more than 1.5 μm. Even after additional 40 hours of growth, a fully coalesced film is not obtained. The center image shows the impinging of the islands forming large boundaries between each other, however, with a high number of pinholes.

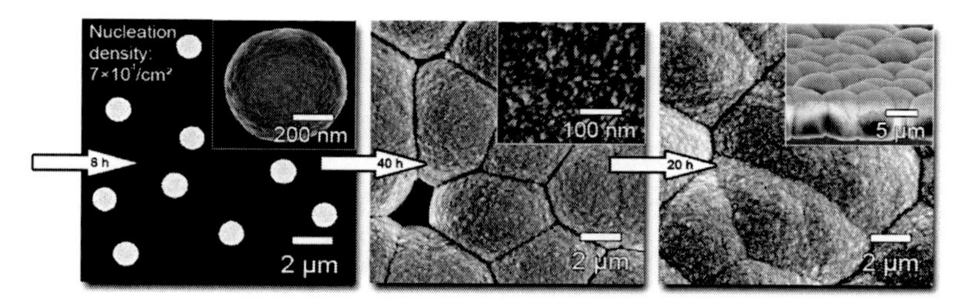

Figure 2.1 NCD nucleation and early-stage growth on an untreated silicon substrate. *Abbreviation*: NCD, nanocrystalline diamond.

Since the growth conditions prevent the growth of large diamond grains, the nanostructure still consists of crystallites in the range of 10 nm (see high-resolution SEM [HRSEM]). Eventually, after an additional 20 hours a fully closed film is achieved. The resulting surface morphology displays the significant impact of the initial high nucleation density. The minimum thickness of a continuous film is in the range of 8–10 μm, having a surface roughness of several hundreds of nanometers. The cross-sectional view bares a further setback of a low nucleation density. Large boundaries that are formed at the merging of the cluster islands clearly proceed through the whole film thickness. Such flaws significantly lower the fracture toughness of the film system. This example displays that NCD growth with all its potentially ultrasmooth film surface and high fracture toughness is strongly affected by the initial pretreatment and nucleation density.

Several substrate pretreatment methods—also called "seeding techniques"—have been developed to increase the nucleation density. The main techniques and the typically resulting nucleation densities on silicon substrates are presented in Table 2.1 and will be discussed in the following sections.

Table 2.1 Surface pretreatments and typical surface nucleation densities [4, 5]

	"Seeding technique"	Nucleation density (cm^{-2})
(a)	No pretreatment	10^3–10^5
(b)	Manual scratching	10^6–10^{10}
(c)	BEN	10^8–10^{11}
(d)	Nanoseeding	10^6–10^{11}

Abbreviation: BEN, bias-enhanced nucleation.

2.1.1 Manual Scratching

The traditional and simplest, yet powerful, method (b) is manually scratching the substrate surface with hard powder grains, that is, microsized diamond particles [6]. The abrasive treatment using diamond fundamentally changes the surface in two ways:

(i) Formation of defects such as dangling bonds, grooves, and scratches
(ii) Implementation of diamond fragments (5–50 nm) into the surface

The formation of defects created on the surface by the hard particles results in chemical-active sites that preferentially adsorb diamond precursors because of their high surface energy. In addition, using diamond particles, the small diamond debris embedded in the silicon surface already offers the diamond structure; hence homoepitaxial growth around the primary nuclei will be favored.

Figure 2.2 presents the scratching method using facetted diamond particles with a size in the range of 50–500 nm and the resulting surface quality. The silicon wafers are polished with a suspension of diamond powder and isopropyl alcohol for approximately 10 minutes. The scratching method is simple and features nucleation densities typically in the range of 10^6–10^{10} cm^{-2}. The right picture and the inset illustrate the proposed effect of the surface scratches and randomly distributed diamond crystals on the surface. A relatively high density is observed along scratches and grooves. The resulting nucleation density here is high, in the range of 10^{10} cm^{-2}, but the pictures clearly bear significant disadvantages.

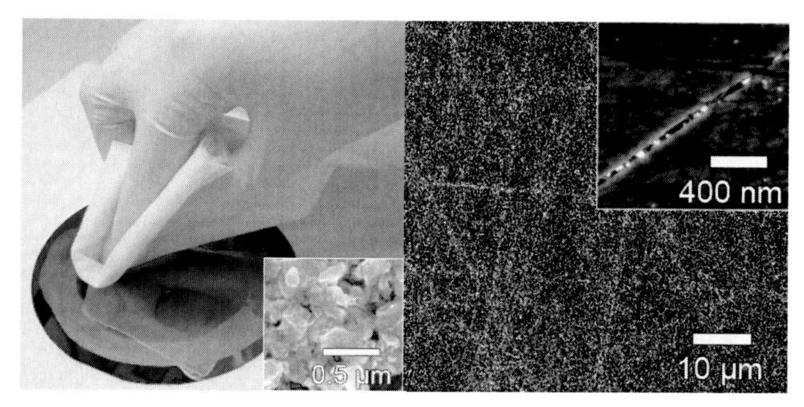

Figure 2.2 (Left) The manual scratching method on a silicon wafer (inset: diamond powder); (right) a manually "seeded" silicon surface.

Scratching with hard powders causes considerable damage to the surface, such as long scratches und grooves. Especially in applications like electronics, optics, and tribology, such surface defects might lead to undefined and undesirable effects. Large grooves and scratches on the surface already predefine surface morphology and can lead to an inhomogeneous roughness distribution. From the industrial aspect, manual scratching shows poor reproducibility, no capability of serial production, and a limitation of substrate geometries. In terms of wear-resistant coatings for three-dimensional (3D) components (e.g., microparts) the manual scratching technique cannot be accomplished.

2.1.2 Bias-Enhanced Nucleation

A further and widely used nucleation technique, first introduced by Yugo et al. [7], is an in situ pretreatment applied prior to diamond growth inside the CVD chamber. The so-called BEN is based on a bombardment of the substrate by active gas species, yielding nucleation densities in the order of 10^8–10^{10} cm^{-2}. A typically setup of the BEN process inside hot-filament CVD (HFCVD) [8] is presented in Fig. 2.3. In general, the gas conditions during BEN are similar to those present in the normal growth process and consist of a mixture of hydrogen (H_2) and methane (CH_4). A positive bias (+40 V) applied to an additionally installed grid accelerates emitted electrons from the hot filaments toward the grid. Carbon-containing molecules and radicals between the filaments and the grid are hit by the electrons and are thereby ionized. A second potential (−300 V) is applied to the substrate holder, which in turn accelerates the charged cations (C^+, CH_x^+, $C_xH_y^+$) from the filaments toward the substrate. It is supposed that the bombardment of the cations onto the substrate's surface induces several effects:

(i) Impact of the carbon-containing cations leads to carbon saturation, resulting in the formation of a graphitic layer. Such a graphitic sp^2 intermediate layer favors diamond nucleation [9].
(ii) Subplantation of carbon ions into the silicon leads to the formation of an intermediate silicon carbide (SiC) layer [10].
(iii) Ion bombardment of the active carbon species induces high local compressive stress, which promotes sp^2 to sp^3 conversion [11].

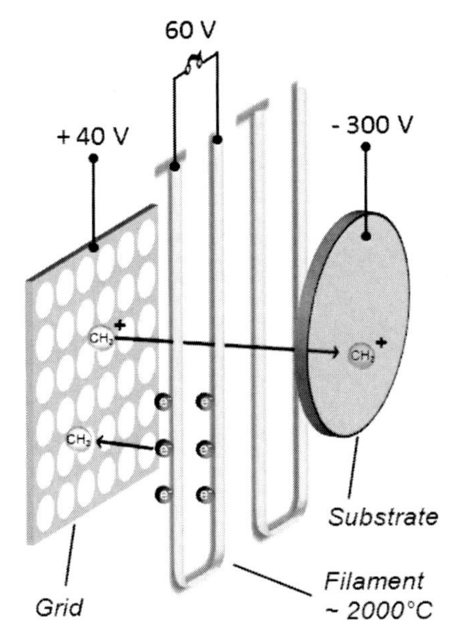

Figure 2.3 Schematic of the BEN procedure.

BEN pretreatment in general induces high and homogeneous nucleation densities; however, a conductive substrate or surface to attract the charged ions is mandatory. The potential for an economic production using the BEN method is not yet evident, considering that the introduced setup, a scale-up, and the employment of 3D objects are rather complex.

2.1.3 Nanoseeding

A promising prenucleation method using a solution with dispersed nanometer-sized diamond particles, also referred to as "nanoseeding," has been recently developed [5, 12, 13]. Nanodiamond crystallites produced from detonation processes [14, 15] exhibiting grains between 5 nm and 20 nm are used in a liquid medium. The substrate or the prestructured wafers can be simply dipped into this solution for a few minutes. Optionally, for increasing the mixing efficiency and to break off agglomerated particles, the process can be performed in an ultrasonic bath. The nanoparticles will be spread homogeneously over the substrate and adhere strongly on the surface owing to electrostatic forces.

This new approach offers great advantages over conventional treatments. It can be applied to most types of substrate material, whereas BEN is limited to conductive surfaces. Regardless of the geometry of the substrates, a highly reproducible and homogeneous nucleation density in all three dimensions is expected.

Representative results from seeding experiments are presented in the following figures. The as-seeded sample (Fig. 2.4) shows a highly homogeneous and dense covering of diamond nanoparticles. In the pictures with higher magnification the distinct agglomeration of the primary nanodiamonds to flakelike clusters (10–100 nm in size) is detected. Since only low-power ultrasonication (25 V) is applied, an implementing of particles on the surface is not expected; thus electrostatic forces are mainly responsible for the adhesion of these diamond particles on the surface.

The formation of NCD crystals at the early stages of diamond deposition is illustrated in Fig. 2.5. After a short growth step of five minutes, the primary diamond seeds are overgrown by nanosized secondary crystallites. The formation of nanometric grains at the active sites is clearly seen in the HRSEM pictures. Since the deposition conditions favor a high rate of secondary nucleation, the newly formed diamond crystals have a maximum size of 10 nm. A homogeneous nucleation density in the order of 10^{11} cm^{-2} over the 3-inch wafer and over a batch of several wafers could be obtained.

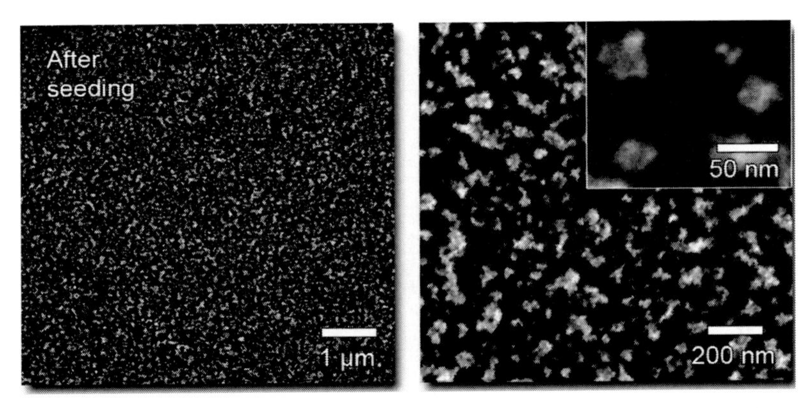

Figure 2.4 Primary nanodiamond particles after nanoseeding pretreatment.

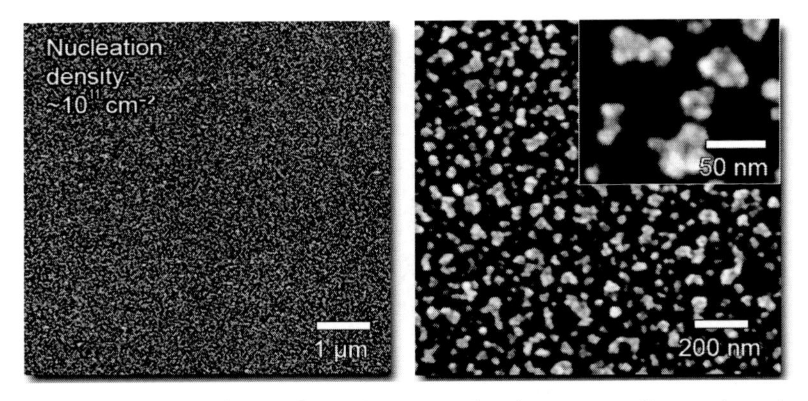

Figure 2.5 NCD cluster formation around primary nanodiamond seeds after a 5 min growth.

The prenucleation techniques using dispersed nanodiamond result in a high and uniform nucleation density in the range of 10^{11} cm^{-2}. Moreover, this method has also shown to be an excellent choice for the seeding of 3D tools and structures and can be easily upscaled for industrial application where high and fast throughput is necessary.

2.2 CVD Techniques for Diamond Growth

2.2.1 Principles of CVD

The basic concept of CVD diamond growth is illustrated in Fig. 2.6. A feed gas, typically composed of H_2 and CH_4, needs to be activated in order to start the diamond formation process. Three techniques with different gas activation techniques are well established in CVD diamond growth. In microwave plasma CVD (MWPCVD) the energy is supplied by electric discharge, generating a plasma. Combustion CVD (CCVD) uses an oxyacetylene torch, and in HFCVD, the gas is thermally activated by hot filaments. Whereas the first two methods (MWPCVD and HFCVD) are widely used for NCD deposition, the combustion CVD technique, which is known for its high growth rates, is only used to produce polycrystalline diamond.

The corresponding activation source causes the molecules to fragment into reactive species such as atomic hydrogen (H) and hydrocarbon (e.g., CH_3) radicals. Furthermore, collisions of atomic hydrogen with hydrocarbon additionally lead to CH_3 radicals.

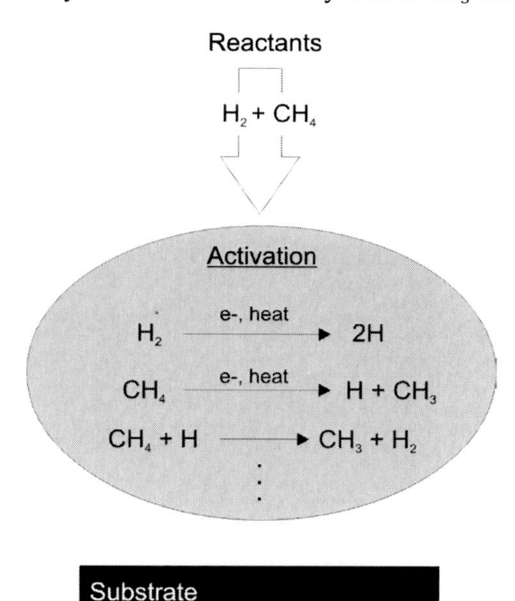

Figure 2.6 Schematic illustration of the CVD diamond process (Reprinted from Ref. 18, by permission of the Royal Society, copyright 1993).

The process after activation can be divided into several steps, which are illustrated in Fig. 2.7 [19]. Any diamond CVD growth starts when individual carbon atoms on the substrate's surface begin to form sp^3 tetrahedral lattices and, thus, the first nuclei. Since nondiamond substrates initially do not promote such a template for the tetrahedral structure, carbon atoms reaching the surface are immediately etched by the highly reactive atomic hydrogen. Such a template can be obtained by using either diamond substrates or particular substrates pretreatments (see section 2.1). The reaction processes presented here are based on an already existing diamond lattice on the surface.

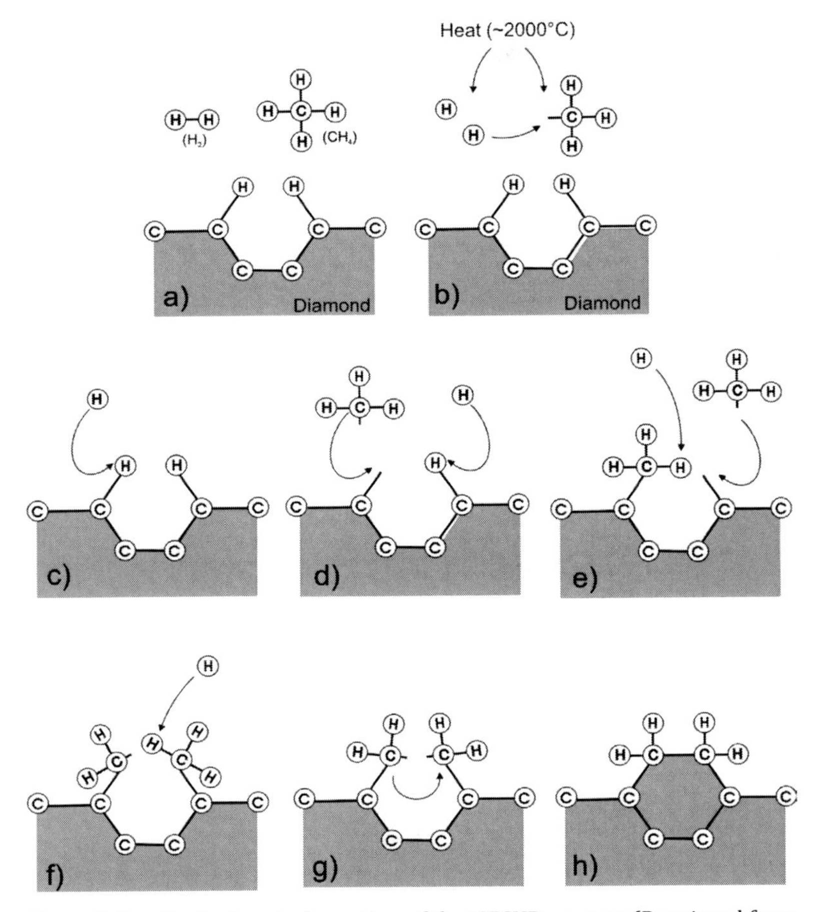

Figure 2.7 Basic chemical reaction of the HFCVD process (Reprinted from Ref. 19, by permission of the Royal Society, copyright 2000).

After fragmentation of the molecules into reactive species (H and CH_3), the atomic hydrogen generates active sites on the hydrogenated diamond surface (Fig. 2.7b,c). Since the gas phase mainly contains active hydrogen atoms, the most likely event is the recombination of the free site to H_2. Sometimes, however, instead of a H atom, a CH_3 radical reacts with the free site, adding a carbon to the lattice (Fig. 2.7d,e). This procedure of adding hydrocarbons continues on the neighboring sites and further hydrogen abstraction on the already attached CH_3 radical takes place (Fig. 2.7e,f). Finally, the adjacent free sites can form C–C bonds, adding two more atoms to the diamond lattice (Fig. 2.7g,h). Consequentially, the diamond formation is based on a stepwise addition of carbon atoms to an existing diamond lattice, catalyzed by the presence of excessive atomic hydrogen. Furthermore, this hydrogen is responsible for preventing the formation of graphite, since atomic hydrogen etches sp^2-bonded carbon (graphite) 20–30 times faster than sp^3-bonded carbon (diamond).

The basic and simplified reaction processes shown here are highly influenced by parameters such as gas mixture, gas pressure, temperature, reactor geometry, and activation source setup, which directly contribute to the diamond quality, grain size, and growth rate.

2.2.2 Hot-Filament CVD

HFCVD, first introduced by Matsumoto et al. [20, 21] in the early 1980s, is the mostly used technique for large-scale deposition of a polycrystalline diamond film. The possibility of coating large areas having 3D structures makes it especially favorable for industrial integration [22, 23]. A sketch of a typical HFCVD reactor with a vertical filament arrangement is presented in Fig. 2.8. The process uses a vacuum chamber (10^{-2}–10^{-4} mbar), and typical deposition conditions for microcrystalline diamond (MCD) growth are 1% methane in 99% hydrogen as the source gas, a filament temperature of 2,000°C, substrate temperature between 500°C and 1,000°C, and gas pressures in the range of 2–20 mbar.

The energy for the activation of the gas phase is provided by the high temperature of the filaments, which can be placed on either side of the substrates, allowing 3D deposition. These filaments are typically made of a refractory material such as molybdenum (Mo), tungsten (W), tantalum (Ta), or rhenium (Re). During the deposition process, the filaments are electrically heated up to 1,800–2,200°C, supplying energy for hydrogen dissociation. Filaments made of refractory metals require a carburization step before they can be effectively used for diamond deposition. Since refractory metals are good carbide formers, an extensive carburization of the filament is unavoidable [24]. As electrical resistivity of subcarbides and carbides is different compared to that of pure metal, a carburization step preceding the actual deposition helps to stabilize the filament. A negative effect of this carbide formation is an increase in the brittleness of the filaments.

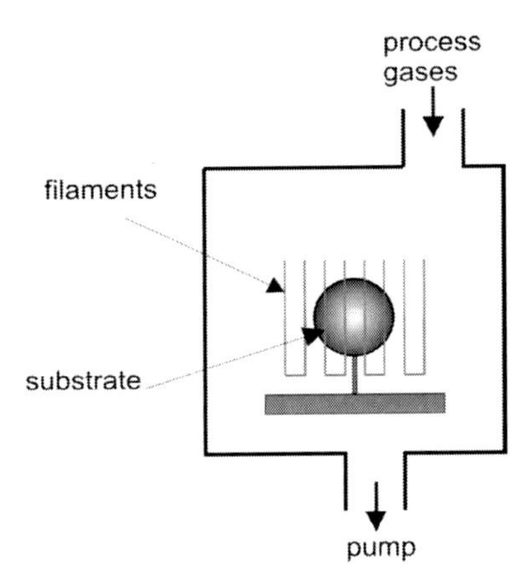

Figure 2.8 Schematic setup of HFCVD.

The primary role of the filaments is to provide enough energy to dissociate diatomic hydrogen into hydrogen atoms via the reaction H_2 + 436 kJ/mol $\leftrightarrow$ 2 H_0 [25]. The secondary role of the filaments is heating the substrate to the deposition temperature for diamond growth. In some setups, the substrate is additionally either heated or cooled. The growth rate strongly depends on the filament temperature, filament arrangement, and substrate temperature. Growth rates using HFCVD for NCD films are typically low and can be found in the range of 10 nm up to 1 μm per hour. A great advantage of this highly cost-effective and simple setup is the ease of the scale-up. At present, a homogeneous and 3D coating of several 6-inch wafers is available.

2.2.3 Microwave Plasma CVD

Along with HFCVD, MWPCVD is an often-used technique to produce NCD. A typical MWPCVD setup is illustrated in Fig. 2.9 [19]. Here the activation of the gas mixture is performed using a glow discharge. This plasma is created by microwave power coupled into the chamber through windows. The energy transfer of the electron–gas collisions leads to fragmentation of the molecules into reactive species. The most-often-used alternating current (AC) frequency of the electric field is 2.45 GHz; sometimes a direct current (DC) plasma is used [19]. The substrate temperature is typically controlled using a heating plate to a value of about 800°C. A potential drawback of all plasma deposition techniques is the limitation to flat substrates. The deposition of 3D substrates is hampered by the inhomogeneous field strength density, especially at the edges, leading to an inhomogeneous film thickness with overaccentuated edges. A typical advantage of MWPCVD is the growth rate, which can be as high as 3 μm/h [26, 27].

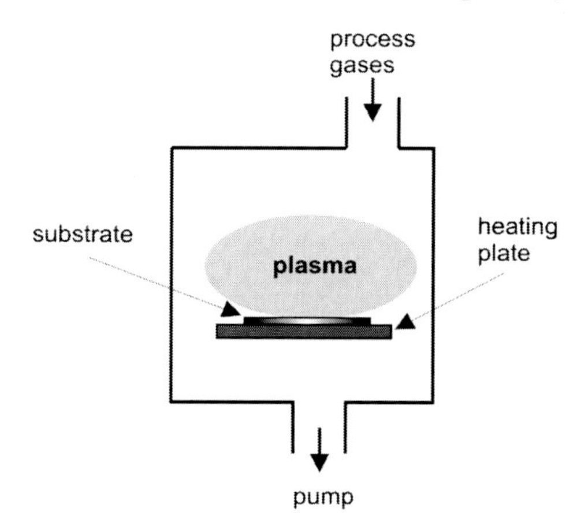

Figure 2.9 Schematic setup of MWPCVD.

2.3 Microcrystalline vs. Nanocrystalline Diamond

In contrast to the HPHT technique, CVD diamond growth results in polycrystalline films. Depending on the distinct growth parameters, the coatings obtained have different microstructures. Nowadays, the crystal size can be controlled from several micrometers down to the nanometer range. Regardless of the size, single grains are separated from each other by grain boundaries, which usually consist of nondiamond components such as graphite or C–H bonds with sp^2- and sp^3-hybridized carbon [28–32]. Hence the films can be viewed as diamond crystals embedded in a nondiamond matrix, and consequently, any variation in crystallite size has a significant influence on the total amount of atoms associated with grain boundaries. Accordingly, the properties can differ depending on the microstructure. Polycrystalline diamond films are commonly classified into three categories according to their crystal sizes as:

(i) MCD
(ii) NCD
(iii) Ultrananocrystalline diamond (UNCD)

Each category offers distinct properties that are owed to, controlled by, and dependent on the grain size. The microstructural differences of the films are schematically illustrated in Fig. 2.10.

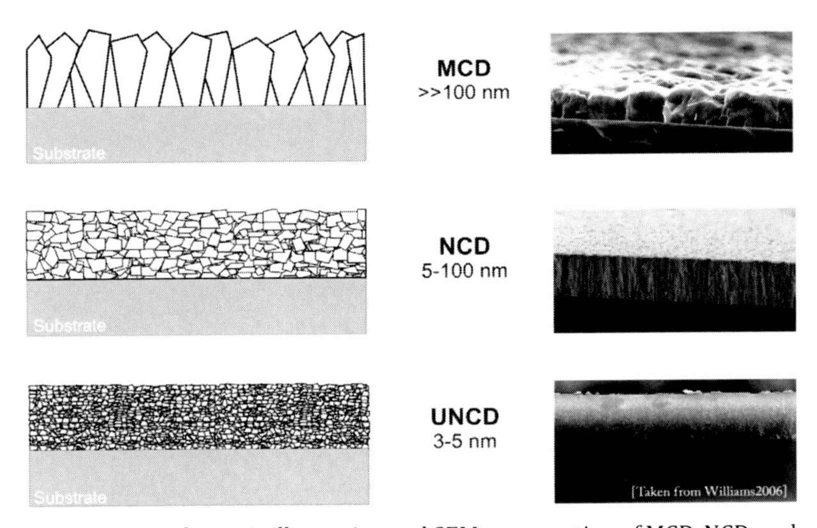

Figure 2.10 Schematic illustration and SEM cross section of MCD, NCD, and UNCD films.

2.3.1 Microcrystalline Diamond Films

MCD films can be seen as the conventional diamond coating, developed about 45 years ago [1–3]. As the name implies, the film consists of large and facetted crystals usually several micrometers in size. A distinct characteristic of MCD films is the nonuniform and anisotropic evolution of the crystallites [33, 34]. As displayed in Fig. 2.11, MCD films typically develop a fiberlike grain texture with increasing crystallite size as the film thickness increases [35]. This texture evolution—sometimes called "columnar growth"—can be described using the "van der Drift growth selection mechanism" [36] and is extensively discussed in the literature [33, 34, 37, 38]. It considers the competitive growth between crystallites having different texture orientations and growth velocities. In the early beginning of the growth phase, the crystallites are randomly orientated. However, with increasing processing time, crystallites having the most rapid growth direction perpendicular to the substrate (i.e., for diamond the {110} direction) will gradually overgrow their neighboring slowly growing grains and hence develop a columnar structure with a preferential orientation {110} normal to the substrate [33, 38]. The crystallites of MCD films are typically equal to the film thickness.

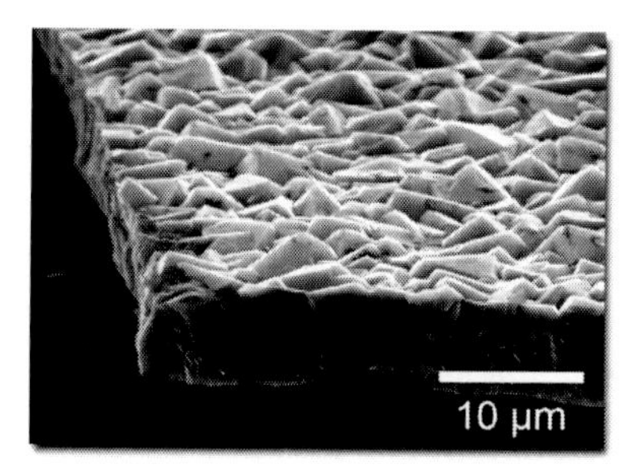

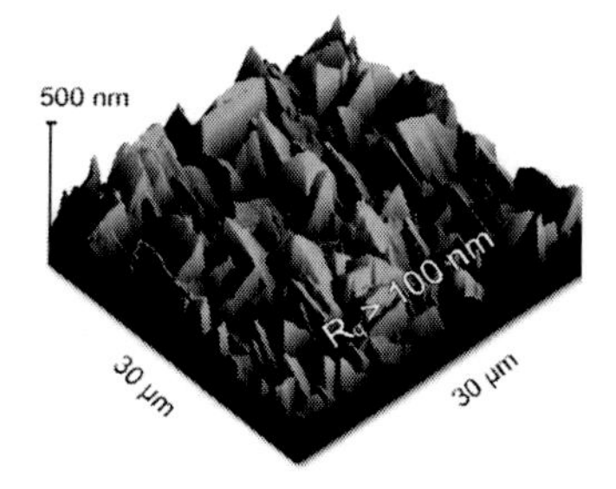

Figure 2.11 SEM and AFM images of an MCD film. *Abbreviation*: AFM, atomic force microscopy.

This evolution phenomenon of MCD films describing anisotropy growth has an essential influence on the material properties. The most specific characteristic of MCD is the high surface roughness, originating from the large and facetted crystallite, which is typically proportional to the film thickness [35, 39, 40]. Roughness values ranging from hundreds of nanometers up to several microns are reported (see Fig. 2.11). Consequently, the originally promising tribomaterial diamond shows relative high friction and poor wear resistance in the microcrystalline state, limiting potential applications [41–44].

Due to the large crystallites in MCD, the mechanical, thermal, optical, and electric properties are very similar to that of single crystalline diamond. The small and negligible amount of grain boundary–related material (<1 %) is reflected in the mass density, which is similar to natural diamond. However, here the implications of the anisotropic microstructure, that is, the increasing grain size with film thickness and the texture evolution, are noticeable. Such a texture can lead to significantly different elastic coefficients, depending on the nature of the preferred orientation [45]. Due to the polycrystalline nature, the hardness in MCD can slightly vary between 80 GPa and 100 GPa, which can be attributed to the orientation of the individual crystals and grain boundary–related weak bonds [46]. Since the overall bonding environment is dominated by sp^3-bonded carbon, the hardness as well as Young's modulus are very similar to those of single crystals [47, 48].

Despite its extreme hardness and stiffness, diamond is very brittle. However, it is still surpassing the most common materials (Si, SiC, SiO_2) used for microelectromechanical systems (MEMS). Especially unfavorable of MCD films is the columnar growth of the grains, leading to a predefined crack path through the film along the grain boundaries. The fracture strength of MCD is reported to be between 0.5 GPa and 1.8 GPa [49, 50].

Furthermore, both the thermal conductivity and the electrical conductivity strongly depend on the crystallographic orientation [51]. Along the film growth direction, phonons can more or less propagate unhindered through the film, whereas along the in-plane direction the phonons are scattered at the grain boundaries, resulting in lower thermal conductivity [52, 53]. While along the growth directions values of about 2,200 W/mK are reported (cf. single crystalline diamond 2,500 W/mK), the in-plane direction offers only about half of the thermal conductivity, which is, however, still an excellent value [54]. The linear thermal expansion coefficient of a large-grained film (~500 nm) is very close to that of natural diamond.

The electrical properties are likewise highly sensitive to grain boundaries. These grain boundaries generally have a high density of dislocations, dangling bonds, or graphitic phases and, therefore, strongly influence the electrical transport properties [55]. An anisotropic grain boundary distribution along the growth directions leads to direction-dependent electrical resistivity in the diamond coating. Typical resistivity varies from 10^2 Ω·cm to 10^8 Ω·cm compared to that of single crystalline diamond (10^{20} Ω·cm) [56–58].

Furthermore, in application of MCD coatings, such as in thermal spreaders or electric devices, the rough surface and the resulting insufficient thermal or electrical contact with its counterparts make it necessary to introduce a surface-polishing step necessary [59].

Diamond is well known for having excellent tribological properties. In general, single crystalline diamond-on-diamond contact sliding in a dry and ambient environment reveals coefficients of friction (COF) in the range of 0.01–0.05 [60]. However, many factors can adversely affect friction and wear behavior. Among others, high surface roughness particularly downgrades the friction coefficient and wear performance. Figure 2.12 shows a collection of friction values as a function of roughness, ranging from 0.03 (for ultrasmooth nanocrystalline films or polished films) up to 0.5 (for very rough microcrystalline films) [60].

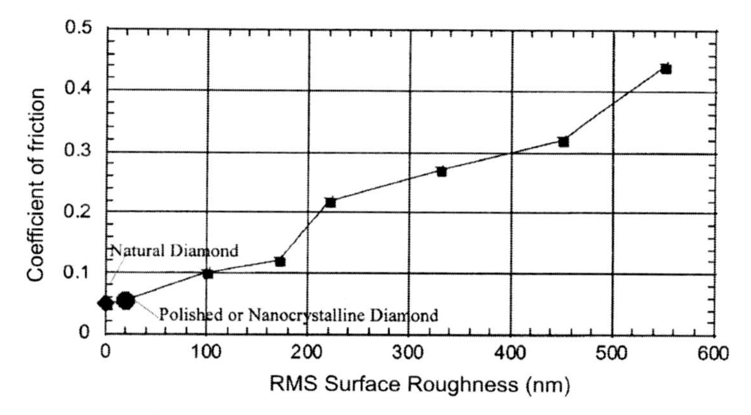

Figure 2.12 Relationship between surface roughness and friction coefficients of CVD diamond films (Republished with permission of Taylor & Francis Group LLC - Books, from Ref. [60]; permission conveyed through Copyright Clearance Center, Inc.).

The wear of MCD films shows a similar trend. When a rough diamond surface having larger protruding grains and asperities slides against a homologous counterpart, considerable fragmentation, plowing, and smoothing are observed. As displayed in Fig. 2.13, the sliding of homologous MCD surfaces typically result in excessive wear rates (10^{-5}–10^{-6} mm^3 N^{-1} m^{-1}) [61].

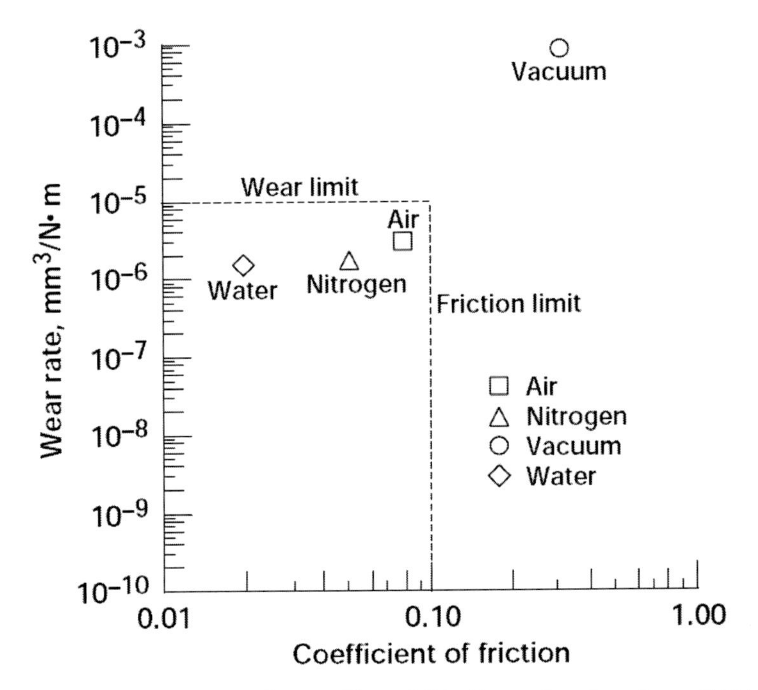

Figure 2.13 Wear rate of MCD films. (This figure is taken from NASA/TM-1999-209189 (Ref. [61]) and used with permission of NASA). *Abbreviation*: NASA, National Aeronautics Space Administration.

Some of the mentioned drawbacks and obstacles of MCD films, especially for tribological and optical applications, can be solved by planarization of the rough diamond surface [59, 62]. However, since diamond is already the hardest-known material, mechanical processing of MCD can be only done by the use of diamond powder and is difficult and time consuming. The same is valid for previously developed polishing methods such as chemical-assisted mechanical polishing, thermochemical polishing, plasma etching, and laser ablation, which show relatively low polishing rates [63, 64]. Even after an extensive polishing process, the required roughness tolerances (typically a few tens of nanometers) for most of the electronic, optical, and tribological applications are, in general, not obtained [65, 66].

In summary, MCD films show properties close to single crystalline diamond; however, distinct microstructural changes with film thickness, such as grain size, grain shape, and grain boundary density, result in anisotropic properties along the growth direction, and most notably the relatively high surface roughness significantly limits the promising employment of MCD in applications. Furthermore, these MCD films exhibit high residual compressive stress and large stress gradients along the growth direction, which typically leads to delamination from the substrate and film bending and reduces the fracture toughness.

2.3.2 (Ultra-)Nanocrystalline Diamond Films

Polycrystalline diamond films having grain sizes less than 100 nm are typically classified as NCD films [30, 67, 68]. A subcategory is UNCD, which was firstly developed by Argonne National Labs, having a crystal size of 3–5 nm and atomically abrupt (~0.5 nm) high-energy grain boundaries [69, 70].

The most mentionable attribute of both categories is that the microstructure is independent of film thickness, that is, the film does not grow in the standard van der Drift regime. Since the classification of NCD films includes UNCD and the differentiation is not well defined in the literature, here only the term "NCD" will be used even for grain sizes less than 10 nm.

An attractive feature of NCD films is the low surface roughness, which strongly depends on the grain size [71, 72]. Previous work showed that the roughness of a diamond film (thickness > 1 μm) can be as low as 10–20 nm [67, 73–77].

As a result of the smooth surface, the tribological properties of NCD films are exceptional. Ball-on-disk experiments at ambient conditions using homologous contacts showed extremely low steady-state friction coefficients as low as 0.01. The positive effect of the small grain size combined with low roughness on the friction is demonstrated in Fig. 2.14. All three coatings show a typical friction evolution of polycrystalline diamond films. After initial high friction values (see inset), also defined as "running in", the curves are decreasing, reaching a steady-state friction coefficient between 0.02 and 0.01 [78]. Regardless of the surface morphology, the three samples reach extreme low friction values. The smoother the initial surface, the faster the plowing event completed, and a low steady-state friction is possible.

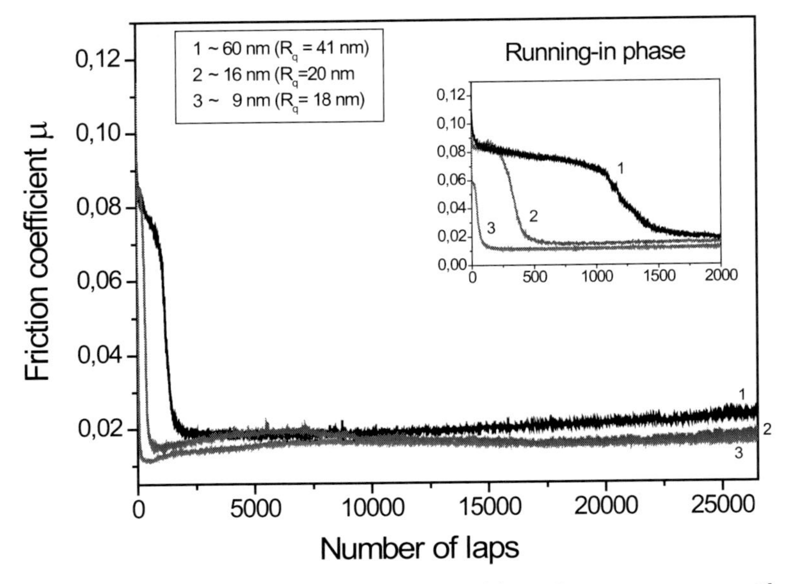

Figure 2.14 Friction coefficient evolution of homologous contacts with different grain sizes under ambient atmosphere. Inset: Running-in behavior [78].

Similar grain size effects can be observed in terms of wear resistance. The residual surface roughness, asperities, or protruding grain clusters on both the balls and the disks leads to initial high contact pressures. With time and sliding revolutions, a smooth plateau on the ball is generated, and on the disk the asperities are chipped and polished away. Consequentially, the smoother the diamond coating, the higher the wear resistance. Taking the values from Fig. 2.15, a grain size reduction from 60 nm to less than 10 nm enhances the wear resistance by more than 140 times.

The relative small dimension of the crystallites in NCD films and, hence, the large volume fraction of grain boundaries significantly alter physical, chemical, and mechanical properties. Whereas in MCD the properties are only little influenced by the grain boundaries, the impact of the disordered interface in NCD is considerably larger. Having, for example, a crystal size of 10 nm, the nondiamond fraction can be in the range of 10% to 30% (depending on the grain boundary thickness). The high nondiamond content is also manifested in the mass density of the coating, which is significantly lower in case of NCD (grain size 10–20 nm), in the range of 3.00–2.75 g/cm^3 [79, 80]. Given that the bulk bonding environment is still solely based on sp^3- and sp^2-bonded carbon and/or hydrogen-terminated sp^3 carbon phases, the great biocompatibility is still sustained even in the nanocrystalline formation [81–83]. However, these hydrogen-/sp^2-bonded carbons, which are mainly found in the grain boundaries, have weaker bond strength compared to diamond [28–32]. Consequently, this weak-bonded atom concentration leads to a decrease in the bulk mechanical properties such as hardness and Young's modulus. Reported hardness values of NCD films can be in the range of 45 GPa to 90 GPa [84], and Young's modulus can vary between 517 GPa and 1,120 GPa [85].

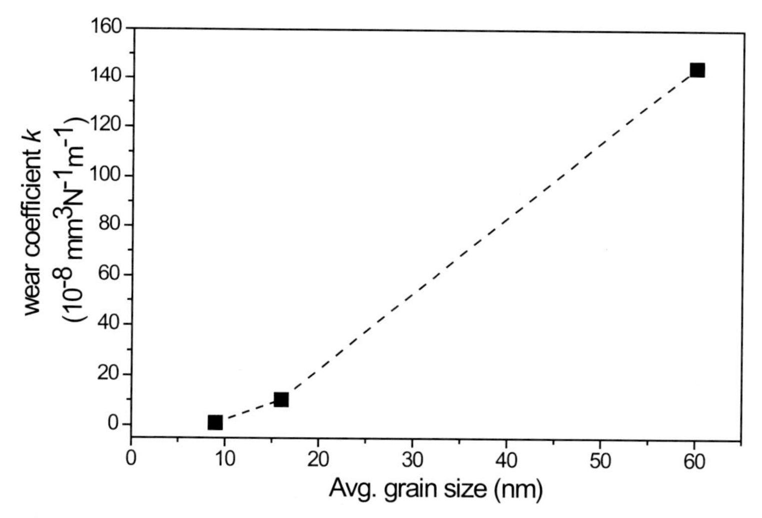

Figure 2.15 Comparison of the wear coefficients (at the same nominal contact pressures) as a function of grain size [78].

A benefit of the high and homogeneous grain boundary density can be seen in the fracture strength of NCD films. With decreasing grain size, the fracture resistance is increasing up to more than double that of MCD. Previously published results showed fracture strength (Weibull approach; stress that would result in 63% of the specimens to fail) of NCD/UNCD films between 2.95 GPa and 4.2 GPa [86–88].

The thermal conductivity of NCD films is mainly determined by the grain boundary width and density [89, 90]. Since in NCD the microstructure is isotropic, that is, grains are equisized in all three dimensions, all phonons are scattered at the grain boundaries. The grain boundary–related atoms will typically not be bonded across the interface and are thus no significant paths for thermal transport. Reported values of k are strongly linked with grain size and typically range from 1.2 W/mK to 26 W/mK [26, 90, 91]. Such low values (~2,000% lower than single crystalline diamond) for the very small grain size is comparable, for example, to glass; thus, it is of little use for thermal management.

Little information about the coefficient of thermal expansion (CTE) of NCD/UNCD films is available in the literature. However, first studies clearly showed that the CTE of NCD films is approximately about a factor of two higher than the one for MCD [92]. Given that the CTE is close to that of silicon, it provides good adhesion of NCD films on silicon.

The electrical properties of NCD films can be controlled by in situ doping with boron or nitrogen. Since the growth of NCD is dominated by intentionally incorporating defects and the grain boundaries are, therefore, decorated with nondiamond components, an intrinsic NCD does not really exist. However, the electric properties are directly connected to the quality and phase purity of the diamond films. While in NCD films the grain boundary density is very large, these nondiamond components distinctly affect the electric properties. Yet, high incorporation of hydrogen into the grain boundaries satisfying sp^2 carbon dangling bonds results in increased resistivity [26, 93].

The key for NCD growth is to offer a high and constant rate of secondary nucleation through deposition, preventing the formation of large crystals [30, 94]. Renucleation is the process by which the evolution of a crystallite is interrupted and a grain boundary is formed, triggering new grain growth [95]. This can be achieved by altering the growth parameters such as gas mixture, pressure, and substrate temperature in order to provide high defect formation [80]. A simple path for NCD growth is to increase the H_2/CH_4 ratio, reducing the effectiveness of the etching effect of atomic hydrogen, hence promoting defect formation and secondary nucleation [96]. However, the lower hydrogen content and the less effective sp^2 etching mechanism in general cause a high degree of carbon supersaturation on the surface and poor film quality with a high degree of nondiamond carbon [97]. Another approach has been developed by replacing hydrogen with the inert gas argon. Using an Ar/CH_4 gas mixture, diamond films with grains of 3–5 nm in size have been reported [98].

In addition to the growth conditions, it is reported that a further important parameter to obtain NCD films is a high primary nucleation density of diamond nuclei on the substrate [5, 68, 99–

102]. A detailed summary of typical growth conditions using various deposition techniques is provided in Ref. 80.

Typical surface structures of NCD/UNCD films having various morphologies are displayed in Fig. 2.16. Example 1 shows an NCD film having distinct crystal faceting with a rather large distribution of grains ranging from 30 nm to 100 nm [103]. This inhomogeneity is not favorable, since properties such as hardness are locally different and protruding large grains negatively affect tribological properties. A further common surface effect in NCD coatings is the formation of "cauliflower like" or "ballas like" structures (see Fig. 2.16, example 2). These spherical clusters consisting of nanometric crystallites exhibit large surface roughness, high compressive stresses, and high content of nondiamond phases [104, 105]. The formation of the ballas-like structure is assumed to be a combination originating from a high methane concentration, leading to a high rate of secondary nucleation and consequently to a high density of defects such as stacking faults and fivefold twins [106–108] and low grain coalescence from a low primary nucleation density. Examples 3 and 4 are the most preferred microstructures for application, having a homogenous size distribution of grains (<20 nm) with smooth surfaces.

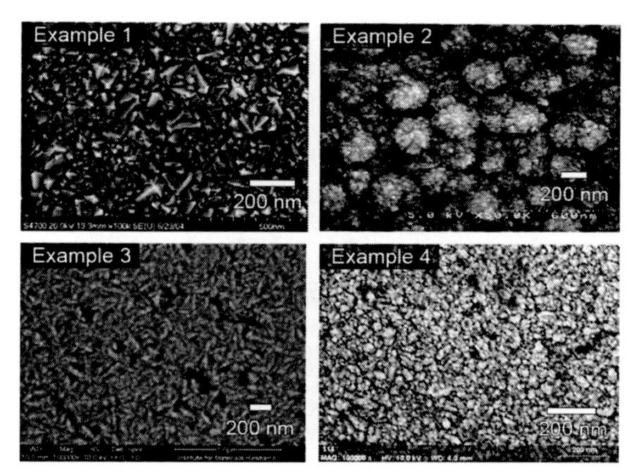

Figure 2.16 Examples of NCD/UNCD films having different microstructures and surface morphologies. (Example 1 reprinted from Ref. [103], copyright 2005, with permission from Elsevier; Example 2 reprinted from Ref. [109], copyright 2005, with permission from Elsevier; Example 3 reprinted from Ref. [110], , copyright 2006, with permission of John Wiley & Sons; Example 4 reprinted from Ref. [111], copyright 2010, with permission from Elsevier.)

In summary, the properties of polycrystalline diamond are mainly dominated by the microstructure, chemical compositions, and grain boundary volume fraction. Since the grain boundary fraction in NCD films can be up to 40%, it is not surprising that such diamond films exhibit considerable changes on many of their physical, chemical, and mechanical properties. Regardless of the declining bulk properties, NCD is a valuable material, especially where hardness, chemical inertness, and outstanding tribological properties are desired.

2.4 Growth Parameter Variation

Diamond growth involves a number of critical deposition parameters, such as gas pressure, gas composition, filament material and filament arrangement, substrate temperature, and reactor geometry. These parameters significantly determine the growth rate, microstructure, and diamond quality. To minimize the grain size, a high rate of secondary nucleation is crucial. Previous research using HFCVD showed the significance of several growth parameters on the microstructure and diamond quality. Schwarz et al. [112], for example, found a direct correlation between gas pressure and grain size. Direct interventions into the gas-phase chemistry by changing the "standard" ratio between H_2 and CH_4 [67, 113, 114] or even introducing additional gases such as oxygen (O_2) [115, 116], nitrogen (N_2) [117–122], argon (Ar) [74, 123], or a combination of Ar and N_2 [124] into the feed gas revealed to have a strong effect on the microstructure of diamond.

In the following two sections, the influence of the gas pressure and gas composition in terms of surface morphology and grain size has been evaluated using the HFCVD technique. The single deposition parameter for the standard process and the varied parameters are listed in Table 2.2.

Table 2.2 Summary of the parameters variation experiment

Parameter	Variation
Gas pressure	3–25 mbar
H_2/CH_4 ratio	Constant
O_2	0–1.60%
N_2	0–7.5%
Filament temperature	Constant
Substrate temperature	Constant

2.4.1 Role of Pressure

Previous research showed a direct influence of gas pressure on diamond growth [112, 124–127]. Gas pressure is an important parameter for CVD diamond growth since gas pressure sensitively changes both the gas-phase fluid dynamics and the gas-phase chemical reaction kinetics among various gas species in a CVD reactor [126]. According to the kinetic theory of gases, the lower the pressure, the larger the mean free path, and the more active species or radicals (atomic H, CH_3, C_2H_2 radical, etc.) survive collisions and reach the substrate surface. Analogously, the higher the pressure, the more active species generated by the filament at a constant filament temperature; however, the mean free path is shorter, and consequently, the collision of hydrogen atoms with each other as well as with other molecules is more frequent. Hence the probability of hydrogen atom recombination and consumption is greater. Furthermore, an increase in gas density leads to an increase of convective dissipation of heat of both the filaments and the substrate. A high collision rate of gas species with the filaments additionally cools the filaments down. This thermal issue in turn lowers the substrate temperature and, finally, leads to a decrease in the growth rate. These contradicting effects of pressure on determining the concentration of active species at the substrates surface obviously need balancing [126].

The details of the "pressure variation" experiments are listed in Table 2.3. All depositions runs took place at a constant H_2/CH_4 ratio with an addition of 0.86% of oxygen. The pressure was varied between 3 mbar and 25 mbar. The tests were performed on 2-inch silicon wafers, manually seeded and deposited for eight hours. It is necessary to mention that both the filament and the substrate temperature was not directly controlled and not monitored during the pressure investigation. Measurement of the gas temperature between the two filament rows showed an increase from 650°C for 3 mbar up to 780°C for 25 mbar gas pressure, which can be associated with an increase in filament temperature or more likely with enhanced convective heat transfer from the filaments. Filament temperature generally influences the ratio of reactive gas species, followed by an influence on the growth rate. A higher filament temperature leads to a higher growth rate. Since in this experiment, no information about the filament temperature is available, the pressure impact on filament and substrate temperature cannot be taken into account.

Table 2.3 Experimental details of "pressure variation"

Parameter	Value/Note
Gas pressure	3–25 mbar
H_2/CH_4 ratio	Constant
O_2	Constant
Deposition time	8 h

The total gas pressure is found to be an essential parameter influencing the diamond grain size. The analysis for the pressure effect on film morphology is presented in Fig. 2.17. A pressure of 25 mbar results in randomly oriented MCD growth with well-defined, large grains (>200 nm). Decreasing the pressure to 15 mbar does not show any significant change in surface morphology. By further decreasing the gas pressure to 8 mbar, the microstructure changes from well facetted to a distribution of large and small crystals showing considerable growth defects. The formation of growth defects such as twinning and secondary nucleation gets more pronounced at a process pressure of 7 mbar and less. The average grain sizes are well below 100 nm, however, still showing a facetted structure. A reduction to 3 mbar shows no further morphological change.

It is reasonable to assume that the change in morphology and the decrease of crystallite size are due to a significant increase of secondary nucleation. As described above, a change in pressure influences the total number of active species generated at the filament, the number of active species reaching the substrate, and the temperature state of the filaments and the substrate. This complex linking of the pressure effects, plus the unknown state of the filament and substrate temperature, makes it challenging to draw explicit conclusions. This result, howsoever, is consistent with previously published findings, showing enhanced secondary nucleation and significantly decreased crystal size by simply reducing gas pressure down to several mbar during CVD deposition [112, 127, 128].

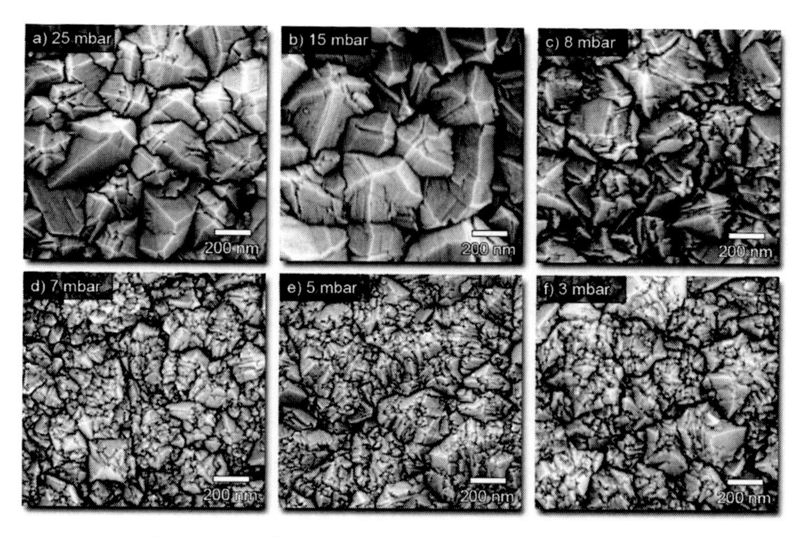

Figure 2.17 Influence of gas pressure on surface morphology and grain size.

In addition to the effect on the microstructure, gas pressure shows a strong impact on the growth rate. Figure 2.18 presents the growth rate as a function of pressure. The growth rate shows a maximum of 216 nm/h at 5 mbar and strongly decreases down to a minimum of 109 nm/h with increasing pressure.

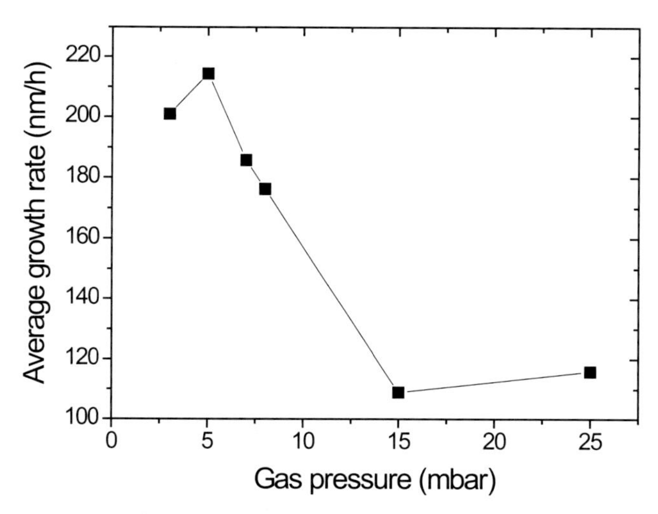

Figure 2.18 Growth rate as a function of gas pressure.

The effect of pressure variation on the morphology and growth rate is manifested. However, the physical and chemical explanations are complex. From the literature the pressure effect on the microstructure is a result of two competing effects, the concentration of the active species generated at the filaments and the mean free path of reactive gas species. If the balance is not right, the diamond growth is evidently disturbed, resulting in smaller grain sizes. A widely accepted explanation is that at high gas pressures and, consequently, with higher gas densities, the dissociation of methane by atomic hydrogen is enhanced due to elevated molecule collisions [97]. According to Hirmke et al. and Schwarz et al., the increased collisions rates lead to a enhanced formation of long-chained carbon molecules and consequently to a decrease of growth species (CH_3, C_2H_2) at the substrate surface. Since the H concentration increases strongly with pressure, a higher etching rate of nondiamond carbon on the substrate is expected. As a result, secondary nucleation is suppressed and the growth rate declines. On the contrary, at low pressure, such as 5 mbar, the concentration of growth species is higher and the produced atomic hydrogen is lower. The concentration of H is not sufficient anymore to suppress growth interruptions and defect formation, which inevitably leads to smaller grain sizes.

2.4.2 Role of the Gas Mixture (CH_4, O_2, N_2)

Selecting the correct gas composition is one of the key factors for successful growth of high-quality diamond films. A typical gas mixture of 1% CH_4 in H_2 results in polycrystalline films with grain sizes in the micron or tens-of-micron range. Changing the hydrogen/methane ratio or even introducing additional gases has promising potential for both enhancing the diamond quality and suppressing the growth of large crystallites.

The mostly used and technologically easiest method to influence the diamond microstructure is simply varying the ratio of the H_2/CH_4 mixture. A low CH_4 concentration ($<0.5\%$) usually leads to large and well-facetted crystallites. The microstructure of these films is highly thickness dependent, resulting in a columnar grain growth with large surface roughness. Increasing the CH_4 concentration reveals a change in surface morphology and crystallite size. It is well accepted in the literature that an increase of the precursor CH_4 in the gas phase and, thus, a decrease of the H_2 concentration highly promote secondary nucleation and finally lead to extreme small crystallites. However, the lower hydrogen content and the less effective sp^2 etching mechanism generally cause a high degree of carbon supersaturation on the surface and poor film quality with a high degree of nondiamond carbon [97]. The grown films have superior surface properties such as low roughness or friction coefficients. However, due to the high sp^2 impurities essential properties such as hardness and Young's modulus are drastically lowered.

The introduction of oxygen into the gas mixture showed beneficial effects on the film quality [19]. In general, the addition of oxygen has a twofold potential, suppressing the growth of nondiamond components and increasing the growth rate. Oxygen dissociates into O atoms and can react with surrounding hydrogen and hydrocarbons into OH, H_2O, CO, and CO_2. It is well known that the highly reactive OH radicals can etch nondiamond components more rapidly than atomic hydrogen by directly oxidizing nondiamond carbon and removing it as CO or CO_2 [115]. Hence, the addition of oxygen into the H_2/CH_4 gas phase of an HFCVD system is a common approach for the production of high-quality MCD films with high growth rates.

The effect of nitrogen in HFCVD is still controversially discussed. The controlled addition of N_2 into the typical H_2/CH_4 gas mixture used for HFCVD showed a significant impact on the gas-phase reaction, diamond microstructure, electronic properties, and growth rate [129–131]. Contrarily, reports showed that nitrogen has no or little effect on the HFCVD process. It is proposed that nitrogen is considered to be an "inert spectator" in an HFCVD environment [132–134]. This, however, has been mainly alluded to in the incorporation of small amounts (ppm) of nitrogen into the gas phase due to vacuum leakage. Larger amounts (0.5–40%) of N_2 into a standard H_2/CH_4 gas phase can result in a strong degradation of crystallinity and diamond phase purity, leading to a smooth nanocrystalline structure [135].

Next, the influence of the oxygen added to the H_2/CH_4 gas mixture on the diamond film properties is discussed. The parameters of the oxygen variation experiments are outlined in Table 2.4. The methane and hydrogen flow as well as the gas pressure are kept constant throughout the experiments. The deposition was carried out on a manually seeded 3-inch Si substrate for eight hours.

Table 2.4 Experimental details of "oxygen variation"

Parameter	Value/Note
Gas pressure	Constant
H_2/CH_4 ratio	Constant
O_2 concentration	0–1.6%
Substrate	3-inch Si wafer
Deposition time	8 h

The oxygen content was varied between 0% and 1.6%. The effect on the surface properties and the microstructure was analyzed by means of SEM (see Fig. 2.19). Both the processes with the highest O_2 concentration (Fig. 2.19a: 1.60%; Fig. 2.19b: 1.31%) clearly evolve large and facetted crystals with a pronounced (111) growth texture. It is assumed that the enhanced etching rate of the hydroxyl radicals (OH) assists diamond growth evidently by compensating nondiamond formation and promotion of secondary nucleation, which is strongly favored through the high CH_4 content. With the reduction of the oxygen concentration, the microstructure evolves from facetted to nonfacetted small grains (Fig. 2.19c–e). The HRSEM pictures in the insets clarify this, showing a decreasing grain size with roundish shapes. From Fig. 2.19f with 0.43% oxygen, a further decrease shows no observable change of surface morphology and microstructure, even in the absence of oxygen. The grain size obtained from the HRSEM images are in the range of 10–20 nm.

The evolution from submicron-sized to nanosized is evidently connected to the oxygen concentration. Since the reactive oxygen accelerates the etch rate of both nondiamond and diamond, it is logical to conclude that the oxygen effect compensates the formation of growth defects formed at high methane concentrations. By reducing the oxygen content, the ratio between defect formation and etching is out of balance, thus promoting renucleation processes and small grain sizes. Investigation of growth rate versus oxygen add-on yields only a minor dependency (not shown here). However, because of inhomogeneous growth over the 3-inch wafer due to setup matter, a clear trend could not be observed.

The addition of nitrogen to interfere with the growth conditions is well applied in MWPCVD. However, in HFCVD the effect is controversially discussed. Four different experiments with varying nitrogen addition were performed (see details in Table 2.5).

Figure 2.19 Influence of oxygen addition on surface morphology and microstructure.

Table 2.5 Experimental details of "nitrogen variation"

Parameter	Value/Note
Gas pressure	Constant
H_2/CH_4 ratio	Constant
N_2 concentration	0.86–7.95%
Substrate	3-inch Si wafer
Deposition time	8 h

The SEM pictures (Fig. 2.20) from the surface of the coatings showed no considerable effect of nitrogen on the microstructure. The grain sizes are comparable to the process using only H_2/CH_4 (Fig. 2.19i), having a roundish shape of 10–20 nm.

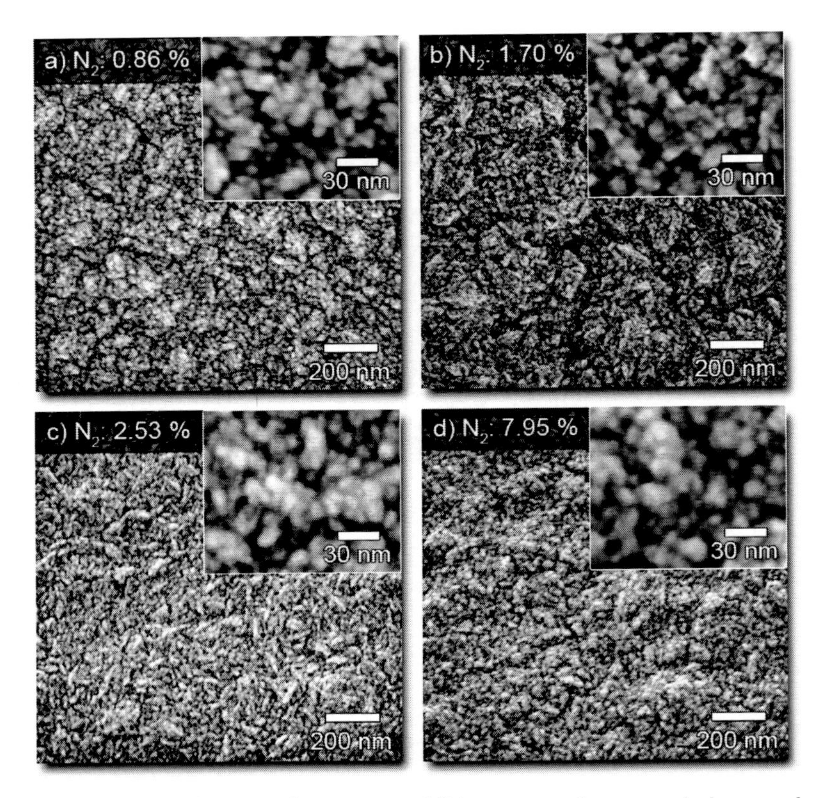

Figure 2.20 Influence of nitrogen addition on surface morphology and microstructure.

The only distinct change was found in the growth rate; with increasing N_2 content, the growth rate was minimized from ~115 nm to ~80 nm. This, however, can be also attributed to a shorter dwell time and higher gas velocity of the H_2 and CH_4 molecules through the chamber, since the pressure was kept constant regardless of the N_2 addition. The contact time between the hydrogen and the hot filament is short, and consequently, the dissociation rate of hydrogen is low, which explains the decreasing growth rate [39].

2.5 Summary

The surface morphology and the microstructure of NCD show a drastic effect on the gas pressure and gas mixture. Diamond growth at a low pressure of ~5 mbar resulted in an evident decrease of defect formation and hence smaller crystallites. As an additional benefit, at 5 mbar the growth rate was maximized. The addition of oxygen to an "abnormal" H_2/CH_4 ratio revealed improved diamond crystallinity. By reducing the oxygen concentration in the feed gas from 1.60% to no O_2, the diamond grain size decreased gradually from submicrometer to nanometer scale. From SEM analysis, a minimum grain size in the order of 10 nm is observed. Adding nitrogen to the H_2/CH_4 phase featured no distinct change of microstructure or surface morphology. It seems that N_2 is indeed, as quoted, an "inert spectator" and only reduces the dwell time of the active species and hence the growth rate.

References

1. Eversole, W.G. (1962). *Synthesis of Diamond*, US Patent 3,030,188.
2. Angus, J.C., Will, H.A., and Stanko, W.S. (1968). Growth of diamond seed crystals by vapor deposition, *J. Appl. Phys.*, **39,** pp. 2915–2922.
3. Deryagin, B.V., and Fedoseev, D.V. (1970). Epitaxial synthesis of diamond in the metastable region, *Russ. Chem. Rev.*, **39,** pp. 783–788.
4. Liu, H., and Dandy, D.S. (1995). Studies on nucleation process in diamond CVD: an overview of recent developments, *Diamond Relat. Mater.*, **4,** pp. 1173–1188.
5. Williams, O.A., Douhéret, O., Daenen, M., Haenen, K., Osawa, E., and Takahashi, M. (2007). Enhanced diamond nucleation on monodispersed nanocrystalline diamond, *Chem. Phys. Lett.*, **445,** pp. 255–258.
6. Mitsuda, Y., Kojima, Y., Yoshida, T., and Akashi, K. (1987). The growth of diamond in microwave plasma under low pressure, *J. Mater. Sci.*, **22,** pp. 1557–1562.
7. Yugo, S., Kanai, T., Kimura, T., and Muto, T. (1991). Generation of diamond nuclei by electric field in plasma chemical vapor deposition, *Appl. Phys. Lett.*, **58,** pp. 1036–1038.
8. Janischowsky, K., Ebert, W., and Kohn, E. (2003). Bias enhanced nucleation of diamond on silicon (100) in a HFCVD system, *Diamond Relat. Mater.*, **12,** pp. 336–339.
9. Robertson, J. (1995). Mechanism of bias-enhanced nucleation and heteroepitaxy of diamond on Si, *Diamond Relat. Mater.*, **4,** pp. 549–552.

10. Kim, Y.K., Han, Y.S., and Lee, J.Y. (1998). The effects of a negative bias on the nucleation of oriented diamond on Si, *Diamond Relat. Mater.*, **7**, pp. 96–105.

11. Ishigaki, N., and Yugo, S. (2000). Mechanism of diamond epitaxial growth on silicon, *Diamond Relat. Mater.*, **9**, pp. 1646–1649.

12. Daenen, M., Williams, O.A., D'Haen, J., Haenen, K., and Nesládek, M. (2006). Seeding, growth and characterization of nanocrystalline diamond films on various substrates, *Phys. Status Solidi (A) Appl. Mater.*, **203**, pp. 3005–3010.

13. Kromka, A., Babchenko, O., Rezek, B., Ledinsky, M., Hruska, K., Potmesil, J., and Vanecek, M. (2009). Simplified procedure for patterned growth of nanocrystalline diamond micro-structures, *Thin Solid Films*, **518**, pp. 343–347.

14. Krüger, A., Kataoka, F., Ozawa, M., Fujino, T., Suzuki, Y., Aleksenskii, A.E., Vul, A.Y., Osawa, E. (2005). Unusually tight aggregation in detonation nanodiamond: identification and disintegration, *Carbon*, **43**, pp. 1722–1730.

15. Shenderova, O.A., Zhirnov, V.V., and Brenner, D.W. (2002). Carbon Nanostructures, *Rev. Solid State Mater. Sci.*, **27**, pp. 227–356.

16. Bongrain, A., Scorsone, E., Rousseau, L., Lissorgues, G., Gesset, C., Saada, S., and Bergonzo, P. (2009). Selective nucleation in silicon moulds for diamond MEMS fabrication, *J. Micromech. Microeng.*, **19**, pp. 1–7.

17. Shenderova, O.A., Hens, S., and McGuire, G. (2009). Seeding slurries based on detonation nanodiamond in DMSO, *Diamond Relat. Mater.*, **19**, pp. 260–267.

18. Butler, J.E., Woodin, R.L., Brown, L.M., and Fallon, P. (1993). Thin film diamond growth mechanisms, *Philos. Trans. R. Soc. Lond. A: Phys. Eng. Sci.*, **342**, pp. 209–224.

19. May, P.W. (2000). Diamond thin films: a 21st-century material, *Phil. Trans. R. Soc. Lond. A: Math. Phys. Eng. Sci.*, **358**, pp. 473–495.

20. Matsumoto, S., Sato, Y., Tsutsumi, M., and Setaka, N. (1982). Growth of diamond particles from methane-hydrogen gas, *J. Mater. Sci.*, **17**, pp. 3106–3112.

21. Matsumoto, S., Sato, Y., Kamo, M., and Setaka, N. (1982). Vapor deposition of diamond particles from methane, *Jpn. J. Appl. Phys. 2*, **21**(N3), pp. 183–185.

22. Schäfer, L., Höfer, M., and Kröger, R. (2006). The versatility of hot-filament activated chemical vapor deposition, *Thin Solid Films*, **515**, pp. 1017–1024.

23. Neto, V.F., Vaz, R., Ali, N., Oliveira, M.S.A., and Grácio, J. (2008). Diamond coatings on 3D structured steel, *Diamond Relat. Mater.*, **17**, pp. 1424–1428.

24. Zeiler, E., Schwarz, S., Rosiwal, S.M., and Singer, R.F. (2002). Structural changes of tungsten heating filaments during CVD of diamond, *Mater. Sci. Eng., A*, **335**, pp. 236–245.

25. Chin, R.P., Huang, J.Y., Shen, Y.R., Chuang, T.J., and Seki, H. (1996). Interactions of hydrogen and methyl radicals with diamond C(111) studied by sum-frequency vibrational spectroscopy, *Phys. Rev. B*, **54**, pp. 8243–8251.

26. Auciello, O., and Sumant, A.V. (2010). Status review of the science and technology of ultrananocrystalline diamond (UNCD) films and application to multifunctional devices, *Diamond Relat. Mater.*, **19**, pp. 699–718.

27. Tang, C.J., Abe, I., Fernandes, A.J.S., Neto, M.A., Gu, L.P., Pereira, S., Ye, H., Jiang, X.F., and Pinto, J.L. A new regime for high rate growth of nanocrystalline diamond films using high power and $CH_4/H_2/N_2/O_2$ plasma, *Diamond Relat. Mater.*, **20**, pp. 304–309.

28. Ballutaud, D., Jomard, F., Kociniewski, T., Rzepka, E., Girard, H., and Saada, S. (2008). Sp^3/sp^2 character of the carbon and hydrogen configuration in micro- and nanocrystalline diamond *Diamond Relat. Mater.*, **17**(4–5), pp. 451–456.

29. Chromik, R.R., Winfrey, A.L., Lüning, J., Nemanich, R.J., and Wahl, K.J. (2008). Run-in behavior of nanocrystalline diamond coatings studied by in situ tribometry, *Wear*, **265**, pp. 477–489.

30. Gruen, D.M. (1999). Nanocrystalline diamond films, *Annu. Rev. Mater. Sci.*, **29**, pp. 211–259.

31. Birrell, J., Gerbi, J.E., Auciello, O., Gibson, J.M., Gruen, D.M., and Carlisle, J.A., (2003). Bonding structure in nitrogen doped ultrananocrystalline diamond, *J. Appl. Phys.*, **93**, pp. 5606–5612.

32. Michaelson, S., Ternyak, O., Akhvlediani, R., Hoffman, A., Lafosse, A., Azria, R., Williams, O.A., and Gruen, D.M. (2007). Hydrogen concentration and bonding configuration in polycrystalline diamond films: from micro-to nanometric grain size, *J. Appl. Phys.*, **102**, pp. 113516 -113527.

33. Wild, C., Herres, N., and Koidl, P. (1990). Texture formation in polycrystalline diamond films, *J. Appl. Phys.*, **68**, pp. 973–978.

34. Liu, T., Raabe, D., and Mao, W.M. (2010). A review of crystallographic textures in chemical vapor-deposited diamond films, *Sig. Image Video Process.*, **4**, pp. 1–16.

35. Abreu, C.S., Oliveira, F.J., Belmonte, M., Fernandes, A.J.S., Silva, R.F., and Gomes, J.R. (2005). Grain size effect on self-mated CVD diamond dry tribosystems, *Wear*, **259**, pp. 771–778.

36. van der Drift, A. (1967). Evolutionary selection, a principle governing growth orientation in vapour-deposited layers, *Philips Res. Rep.*, **22**, pp. 267–288.

37. Thompson, C.V. (2000). Structure evolution during processing of polycrystalline films, *Annu. Rev. Mater. Sci.*, **30**, pp. 159–190.

38. Clausing, R.E., Heatherly, L., Horton, L.L., Specht, E.D., Begun, G.M., and Wang, Z.L. (1992). Textures and morphologies of chemical vapor deposited (CVD) diamond, *Diamond Relat. Mater.*, **1**, pp. 411–415.

39. Haubner, R., and Lux, B. (1993). Diamond growth by hot-filament chemical vapor deposition: state of the art, *Diamond Relat. Mater.*, **2**, pp. 1277–1294.

40. Williams, O.A., Daenen, M., D'Haen, J., Haenen, K., Maes, J., Moshchalkov, V.V., Nesládek, M., and Gruen, D.M. (2006). Comparison of the growth and properties of ultrananocrystalline diamond and nanocrystalline diamond, *Diamond Relat. Mater.*, **15**, pp. 654–658.

41. Gardos, M.N. (1999). Tribological fundamentals of polycrystalline diamond films, *Surf. Coat. Technol.*, **113**, pp. 183–200.

42. Rats, D., Vandenbulcke, L., Boher, C., and Farges, G. (1997). Tribological study of diamond coatings on titanium alloys, *Surf. Coat. Technol.*, **94–95**, pp. 555–560.

43. Vandenbulcke, L., and De Barros, M.I. (2001). Deposition, structure, mechanical properties and tribological behavior of polycrystalline to smooth fine-grained diamond coatings, *Surf. Coat. Technol.*, **146–147**, pp. 417–424.

44. Schade, A., Rosiwal, S.M., and Singer, R.F. (2007). Influence of surface topography of HF-CVD diamond films on self-mated planar sliding contacts in dry environments, *Surf. Coat. Technol.*, **201**, pp. 6197–6205.

45. Klein, C.A. (1992). Anisotropy of Young's modulus and Poisson's ratio in diamond, *Mater. Res. Bull.*, **27**, pp. 1407–1414.

46. Savvides, N., and Bell, T.J. (1992). Microhardness and Young's modulus of diamond and diamondlike carbon films, *J. Appl. Phys.*, **72**(7), pp. 2791–2796.

47. Szuecs, F., Werner, M., Sussmann, R.S., Pickles, C.S.J., and Fecht, H.J. (1999). Temperature dependence of Young's modulus and degradation of chemical vapor deposited diamond, *J. Appl. Phys.*, **86**, pp. 6010–6017.

48. Chowdhury, S., de Barra, E., and Laugier, M.T. (2004). Study of mechanical properties of CVD diamond on SiC substrates, *Diamond Relat. Mater.*, **13**, pp. 1625–1631.

49. Hollman, P., Alahelisten, A., Olsson, M., and Hogmark, S. (1995). Residual stress, Young's modulus and fracture stress of hot flame deposited diamond, *Thin Solid Films*, **270**, pp. 137–142.

50. Lu, F.X., Jiang, Z., Tang, W.Z., Huang, T.B., and Liu, J.M. Accurate measurement of strength and fracture toughness for miniature-size thick diamond-film samples by three-point bending at constant loading rate, *Diamond Relat. Mater.*, **10**, pp. 770–774.

51. Jin, S., and Mavoori, H. (1998). Processing and properties of CVD diamond for thermal management, *J. Electron. Mater.*, **27**, pp. 1148–1153.

52. Popovich, A., Ralchenko, V., Konov, V., Saveliev, A., Khomich, A., Zharikov, E., and Prokhorov, A.M. (2005). Thermal conductivity of poly- and nanorystalline diamond films grown in microwave plasma, in *II France-Russia Seminar, New Achievements in Material Science* (Moscow, Russia).

53. Wörner, E., Wild, C., Müller-Sebert, W., Locher, R., and Koidl, P. (1996). Thermal conductivity of CVD diamond films: High-precision, temperature-resolved measurements, *Diamond Relat. Mater.*, **5**, pp. 688–692.

54. Malshe, A.P., and Brown, W.D. (2002). Chapter 11, "Diamond heat spreaders and thermal management," in *Diamond Films Handbook*, eds. Asmussen, J., and Reinhard, D.K. (CRC Press, New York).

55. Isberg, J. (2009). "Transport properties of electrons and holes in diamond,"in *CVD Diamond for Electronic Devices and Sensors*, ed. Sussmann, R.S. (Wiley, Wiltshire), pp. 29–48.

56. Muto, Y., Sugino, T., Shirafuji, J., and Kobashi, K. (1991). Electrical conduction in undoped diamond films prepared by chemical vapor deposition, *Appl. Phys. Lett.*, **59**, pp. 843–845.

57. Shiomi, H., Tanabe, K., Nishibayashi, Y., and Fujimori, N. (1990). Epitaxial growth of high quality diamond film by the microwave plasma-assisted chemical-vapor-deposition method, *Jpn. J. App. Phys.*, **29**, pp. 34–40.

58. Kulkarni, A.K., Tey, K., and Rodrigo, H. (1995). Electrical characterization of CVD diamond thin films grown on silicon substrates, *Thin Solid Films*, **270**, pp. 189–193.

59. Malshe, A.P., Park, B.S., Brown, W.D., and Naseem, H.A. (1999). A review of techniques for polishing and planarizing chemically vapor-

deposited (CVD) diamond films and substrates, *Diamond Relat. Mater.*, **8**, pp. 1198–1213.

60. Erdemir, A., and Donnet, C. (2000). "Tribology of diamond, diamond-like carbon and related films," in *Modern Tribology Handbook*, ed. Bhushan, B. (CRC Press, Boca Raton), pp. 819–856.

61. Miyoshi, K.M., Watanabe, M.S., Takeuchi, S., Wu and R.L.C. (1999). Tribological characteristics and applications of superhard coatings: CVD diamond, DLC, and c-BN, *Proc. 1999 Appl. Diamond Conf./Frontier Carbon Technol. Joint Conf.*

62. Hird, J.R., and Field, J.E. (2004). Diamond polishing, *Proc. R. Soc. Lond. A*, **460**, pp. 3547–3568.

63. Tang, C.J., Neves, A.J., Fernandes, A.J.S., Grácio, J., and Ali, N. (2003). A new elegant technique for polishing CVD diamond films, *Diamond Relat. Mater.*, **12**, pp. 1411–1416.

64. Johnson, C.E. (1994). Chemical polishing of diamond, *Surf. Coat. Technol.*, **68–69**(C), pp. 374–377.

65. Bhushan, B., Subramaniam, V.V., Malshe, A., Gupta, B.K., and Ruan, J. (1993). Tribological properties of polished diamond films, *J. Appl. Phys.*, **74**, pp. 4174–4180.

66. Wilks, E.M., and Wilks, J. (1972). The resistance of diamond to abrasion, *J. Phys. D: Appl. Phys.*, **5**, pp. 1902–1919.

67. Gruen, D.M., Pan, X., Krauss, A.R., Liu, S., Luo, J., and Foster, C.M. (1994). Deposition and characterization of nanocrystalline diamond films, *J. Vac. Sci. Technol. A*, **12**, pp. 1491–1495.

68. Erz, R., Doetter, W., Jung, K., and Ehrhardt, H. (1993). Preparation of smooth and nanocrystalline diamond films, *Diamond Relat. Mater.*, **2**, pp. 449–453.

69. Krauss, A.R., Auciello, O., Gruen, D.M., Jayatissa, A., Sumant, A., Tucek, J., Mancini, D.C., Moldovan, N., Erdemir, A., Ersoy, D., Gardos, M.N., Busmann, H.G., Meyer, E.M., and Ding, M.Q. (2001). Ultrananocrystalline diamond thin films for MEMS and moving mechanical assembly devices, *Diamond Relat. Mater.*, **10**, pp. 1952–1961.

70. Huang, W.S., Tran, D.T., Asmussen, J., Grotjohn, T.A., and Reinhard, D. (2006). Synthesis of thick, uniform, smooth ultrananocrystalline diamond films by microwave plasma-assisted chemical vapor deposition, *Diamond Relat. Mater.*, **15**, pp. 341–344.

71. Wiora, M., Bruehne, K., Floeter, A., Gluche, P., Willey, T.M., Kucheyev, S.O., Van Buuren, A.W., Hamza, A.V., Biener, J., and Fecht, H.-J. (2009). Grain size dependent mechanical properties of nanocrystalline

diamond films grown by hot-filament CVD, *Diamond Relat. Mater.*, **18**, pp. 927–930.

72. Oliveira, T.J., and Aarao Reis, F.D.A. (2007). Effects of grains' features in surface roughness scaling, *J. Appl. Phys.*, **101**, pp. 063507–1 - 063507–7.

73. Lee, Y.C., Lin, S.J., Buck, V., Kunze, R., Schmidt, H., Lin, C.Y., Fang, W.L., and Lin, I.N. (2008). Surface acoustic wave properties of natural smooth ultra-nanocrystalline diamond characterized by laser-induced SAW pulse technique, *Diamond Relat. Mater.*, **17**, pp. 446–450.

74. Amaral, M., Fernandes, A.J.S., Vila, M., Oliveira, F.J., and Silva, R.F. (2006). Growth rate improvements in the hot-filament CVD deposition of nanocrystalline diamond, *Diamond Relat. Mater.*, **15**, pp. 1822–1827.

75. Zhang, J., Zimmer, J.W., Howe, R.T., and Maboudian, R. (2008). Characterization of boron-doped micro- and nanocrystalline diamond films deposited by wafer-scale hot filament chemical vapor deposition for MEMS applications, *Diamond Relat. Mater.*, **17**, pp. 23–28.

76. Reinhard, D.K., Grotjohn, T.A., Becker, M., Yaran, M.K., Schuelke, T., and Asmussen, J. (2004). Fabrication and properties of ultranano, nano, and microcrystalline diamond membranes and sheets, *J. Vac. Sci. Technol. B: Microelectron. Nanometer Struct.*, **22**, pp. 2811–2817.

77. Chen, Y.C., Zhong, X.Y., Konicek, A.R., Grierson, D.S., Tai, N.H., Lin, I.N., Kabius, B., Hiller, J.M., Sumant, A.V., Carpick, R.W., and Auciello, O. (2008). Synthesis and characterization of smooth ultrananocrystalline diamond films via low pressure bias-enhanced nucleation and growth, *Appl. Phys. Lett.*, **92**, pp. 133113–133116.

78. Wiora, M. (2013). *Characterization of Nanocrystalline Diamond Coatings for Micro-Mechanical Applications*, Doctoral dissertation, Ulm University, Germany.

79. Popov, C., Kulisch, W., Gibson, P.N., Ceccone, G., and Jelinek, M. (2004). Growth and characterization of nanocrystalline diamond/amorphous carbon composite films prepared by MWCVD, *Diamond Relat. Mater.*, **13**, pp. 1371–1376.

80. Kulisch, W., and Popov, C. (2006). On the growth mechanisms of nanocrystalline diamond films, *Phys. Status Solidi (A) Appl. Mater.*, **203**, pp. 203–219.

81. Schrand, A.M., Dai, L., Schlager, J.J., Hussain, S.M., and Osawa, E. (2007). Differential biocompatibility of carbon nanotubes and nanodiamonds, *Diamond Relat. Mater.*, **16**, pp. 2118–2123.

82. Yu, S.-J., Kang, M.-W., Chang, H.-C., Chen, K.-M., and Yu, Y.-C. (2005). Bright fluorescent nanodiamonds: no photobleaching and low cytotoxicity, *J. Am. Chem. Soc.*, **127**, pp. 17604–17605.

83. Rodil, S.E., Olivares, R., Arzate, H., and Muhl, S. Properties of carbon films and their biocompatibility using in-vitro tests, *Diamond Relat. Mater.*, **12**, pp. 931–937.

84. Catledge, S.A., Borham, J., Vohra, Y.K., Lacefield, W.R., and Lemons, J.E. (2002). Nanoindentation hardness and adhesion investigations of vapor deposited nanostructured diamond films, *J. Appl. Phys.*, **91**, pp. 5347–5352.

85. Philip, J., Hess, P., Feygelson, T., Butler, J.E., Chattopadhyay, S., Chen, K.H., and Chen, L.C. (2003). Elastic, mechanical, and thermal properties of nanocrystalline diamond films, *J. Appl. Phys.*, **93**, pp. 2164–2171.

86. Guillén, F.J.H., Janischowsky, K., Kusterer, J., Ebert, W., and Kohn, E. (2005). Mechanical characterization and stress engineering of nanocrystalline diamonds films for MEMS applications, *Diamond Relat. Mater.*, **14**, pp. 411–415.

87. Espinosa, H.D., Peng, B., Prorok, B.C., Moldovan, N., Auciello, O., Carlisle, J.A., Gruen, D.M., and Mancini, D.C. (2003). Fracture strength of ultrananocrystalline diamond thin films—identification of Weibull parameters, *J. Appl. Phys.*, **94**, pp. 6076–6084.

88. Ramakrishnan, R., Lodes, M.A., Rosiwal, S.M., and Singer, R.F. (2011). Self-supporting nanocrystalline diamond foils: influence of template morphologies on the mechanical properties measured by ball on three balls testing, *Acta Mater.*, **59**(9), pp. 3343–3351.

89. Liu, W.L., Shamsa, M., Calizo, I., Balandin, A.A., Ralchenko, V., Popovich, A., and Saveliev, A. (2006). Thermal conduction in nanocrystalline diamond films: effects of the grain boundary scattering and nitrogen doping, *Appl. Phys. Lett.*, **89**, pp. 171915-3–171915-4.

90. Angadi, M.A., Watanabe, T., Bodapati, A., Xiao, X., Auciello, O., Carlisle, J.A., Eastman, J.A., Keblinski, P., Schelling, P.K., and Phillpot, S.R. (2006). Thermal transport and grain boundary conductance in ultrananocrystalline diamond thin films, *J. Appl. Phys.*, **99**, pp. 114301–114307.

91. Shamsa, M., Ghosh, S., Calizo, I., Ralchenko, V., Popovich, A., and Balandin, A.A. (2008). Thermal conductivity of nitrogenated ultrananocrystalline diamond films on silicon, *J. Appl. Phys.*, **103**, pp. 083538-1–083538-8.

92. Woehrl, N., Hirte, T., Posth, O., and Buck, V. (2009). Investigation of the coefficient of thermal expansion in nanocrystalline diamond films, *Diamond Relat. Mater.*, **18**, pp. 224–228.

93. Liu, C., Xiao, X., Wang, J., Shi, B., Adiga, V.P., Carpick, R.W., Carlisle, J.A., and Auciello, O. (2007). Dielectric properties of hydrogen-incorporated chemical vapor deposited diamond thin films, *J. Appl. Phys.*, **102**, pp. 74115–74122.

94. Hogmark, S., Hollman, P., Alahelisten, A., and Hedenqvist, P. (1996). Direct current bias applied to hot flame diamond deposition produces smooth low friction coatings, *Wear*, **200**, pp. 225–232.

95. Williams, O.A. (2011). Nanocrystalline diamond, *Diamond Relat. Mater.*, **20**, pp. 621–640.

96. Ramos, S.C., Azevedo, A.F., Baldan, M.R., and Ferreira, N.G. (2009). Effect of methane addition on ultrananocrystalline diamond formation: Morphology changes and induced stress, *J. Vac. Sci. Technol. A: Vac. Surf. Films*, **28**, pp. 27–32.

97. Hirmke, J., Hempel, F., Stancu, G.D., Röpcke, J., Rosiwal, S.M., and Singer, R.F. (2006). Gas-phase characterization in diamond hot-filament CVD by infrared tunable diode laser absorption spectroscopy, *Vacuum*, **80**, pp. 967–976.

98. Zhou, D., McCauley, T.G., Qin, L.C., Krauss, A.R., and Gruen, D.M. (1998). Synthesis of nanocrystalline diamond thin films from an Ar-CH_4 microwave plasma, *J. Appl. Phys.*, **83**, pp. 540–543.

99. Popov, C., Favaro, G., Kulisch, W., and Reithmaier, J.P. (2009). Influence of the nucleation density on the structure and mechanical properties of ultrananocrystalline diamond films, *Diamond Relat. Mater.*, **18**, pp. 151–154.

100. Yang, W.B., Lü, F.X., and Cao, Z.X. (2002). Growth of nanocrystalline diamond protective coatings on quartz glass, *J. Appl. Phys.*, **91**, pp. 10068–10073.

101. Bull, S.J., Chalker, P.R., Johnston, C., and Copper, C.V. (1994). Indentation response of diamond thin films, *Diamond Relat. Mater.*, **4**, pp. 43–52.

102. Chen, L.C., Kichambare, P.D., Chen, K.H., Wu, J.J., Yang, J.R., and Lin, S.T. (2001). Growth of highly transparent nanocrystalline diamond films and a spectroscopic study of the growth, *J. Appl. Phys.*, **89**, pp. 753–759.

103. Mortet, V., D'Haen, J., Potmesil, J., Kravets, R., Drbohlav, I., Vorlicek, V., Rosa, J., and Vanecek, M. (2005). Thin nanodiamond membranes and their microstructural, optical and photoelectrical properties, *Diamond Relat. Mater.*, **14**, pp. 393–397.

104. Sharda, T., Umeno, M., Soga, T., and Jimbo, T. (2001). Strong adhesion in nanocrystalline diamond films on silicon substrates, *J. Appl. Phys.*, **89**, pp. 4874–4878.

105. Lifshitz, Y., Meng, X.M., Lee, S.T., Akhveldiany, R., and Hoffman, A. (2004). Visualization of diamond nucleation and growth from energetic species, *Phys. Rev. Lett.*, **93**, pp. 056101-1–056101-4.

106. Jiang, X., and Jia, C.L. (2002). Structure and defects of vapor-phase-grown diamond nanocrystals, *Appl. Phys. Lett.*, **80,** 13, pp. 2269–2271.

107. Zhou, X.T., Li, Q., Meng, F.Y., Bello, I., Lee, C.S., Lee, S.T., and Lifshitz, Y. (2002). Manipulation of the equilibrium between diamond growth and renucleation to form a nanodiamond/amorphous carbon composite, *Appl. Phys. Lett.*, **80,** 18, pp. 3307–3309.

108. Shechtman, D., Feldman, A., Vaudin, M.D., and Hutchison, J.L. (1993). Moire fringe images of twin boundaries in chemical vapor deposited diamond, *Appl. Phys. Lett.*, **62**, pp. 487–489.

109. Subramanian, K., Kang, W.P., Davidson, J.L., and Hofmeister, W.H. (2005). The effect of growth rate control on the morphology of nanocrystalline diamond, *Diamond Relat. Mater.*, **14**, pp. 404–410.

110. Williams, O.A., and Nesládek, M. (2006). Growth and properties of nanocrystalline diamond films, *Phys. Status Solidi (A) Appl. Mater.*, **203**, pp. 3375–3386.

111. Williams, O.A., Kriele, A., Hees, J., Wolfer, M., Müller-Sebert, W., and Nebel, C.E. (2010). High Young's modulus in ultra thin nanocrystalline diamond, *Chem. Phys. Lett.*, **495**, pp. 84–89.

112. Schwarz, S., Rosiwal, S.M., Frank, M., Breidt, D., and Singer, R.F. (2002). Dependence of the growth rate, quality, and morphology of diamond coatings on the pressure during the CVD-process in an industrial hot-filament plant, *Diamond Relat. Mater.*, **11**, pp. 589–595.

113. Azevedo, A.F., Ramos, S.C., Baldan, M.R., and Ferreira, N.G. (2008). Graphitization effects of CH_4 addition on NCD growth by first and second Raman spectra and by X-ray diffraction measurements, *Diamond Relat. Mater.*, **17**, pp. 1137–1142.

114. Rakha, S.A., Yang, S., He, Z., Ahmed, I., Zhu, D., and Gong, J. (2009). Synthesis of thin, diamond films from faceted nanosized crystallites, *Curr. Appl. Phys.*, **9**, pp. 698–702.

115. Kim, Y.K., Jung, J.H., Lee, J.Y., and Ahn, H.J. (1995). The effects of oxygen on diamond synthesis by hot-filament chemical vapor deposition, *J. Mater. Sci: Mater. Electron.*, **6**, pp. 28–33.

116. Kawato, T., and Kondo, K. (1987). Effects of oxygen on CVD diamond synthesis, *Jpn. J. Appl. Phys. 1*, **26**, pp. 1429–1432.

117. Cao, G.Z., Schermer, J.J., Van Enckevort, W.J.P., Elst, W.A.L.M., and Giling, L.J. (1996). Growth of {100} textured diamond films by the addition of nitrogen, *J. Appl. Phys.*, **79**, pp. 1357–1364.

118. Afzal, A., Rego, C.A., Ahmed, W., and Cherry, R.I. (1998). HFCVD diamond grown with added nitrogen: Film characterization and gas-phase composition studies, *Diamond Relat. Mater.*, **7**, pp. 1033–1038.

119. Zhang, G.F., Geng, D.S., and Yang, Z.J. (1999). High nitrogen amounts incorporated diamond films deposited by the addition of nitrogen in a hot-filament CVD system, *Surf. Coat. Technol.*, **122**, pp. 268–272.

120. Yu, Z., Karlsson, U., and Flodström, A. (1999). Influence of oxygen and nitrogen on the growth of hot-filament chemical vapor deposited diamond films, *Thin Solid Films*, **342**, pp. 74–82.

121. Bohr, S., Haubner, R., and Lux, B. (1996). Influence of nitrogen additions on hot-filament chemical vapor deposition of diamond, *Appl. Phys. Lett.*, **68**, pp. 1075–1077.

122. Chih, Y.K., Chueh, Y.L., Chen, C.H., Hwang, J., Chou, L.J., and Kou, C.S. (2006). Nano-scale diamond tips: synthesis in the CH_4/ N_2/ H_2 plasma, *Diamond Relat. Mater.*, **15**, pp. 1246–1249.

123. May, P.W., Smith, J.A., and Mankelevich, Y.A. (2006). Deposition of NCD films using hot filament CVD and Ar/CH_4/ H_2 gas mixtures, *Diamond Relat. Mater.*, **15**, pp. 345–352.

124. Rakha, S.A., Xintai, Z., Zhu, D., and Guojun, Y. (2010). Effects of N_2 addition on nanocrystalline diamond films by HFCVD in Ar/CH_4 gas mixture, *Curr. Appl. Phys.*, **10**, pp. 171–175.

125. Harris, S.J., and Weiner, A.M. (1994). Pressure and temperature effects on the kinetics and quality of diamond films, *J. Appl. Phys.*, **75**, pp. 5026–5032.

126. Yang, S., He, Z., Li, Q., Zhu, D., and Gong, J. (2008). Diamond films with preferred <110> texture by hot filament CVD at low pressure, *Diamond Relat. Mater.*, **17**, pp. 2075–2079.

127. Liang, X., Wang, L., Zhu, H., and Yang, D. (2007). Effect of pressure on nanocrystalline diamond films deposition by hot filament CVD technique from CH_4/H_2 gas mixture, *Surf. Coat. Technol.*, **202**, pp. 261–267.

128. Wang, T., Xin, H.W., Zhang, Z.M., Dai, Y.B., and Shen, H.S. (2004). The fabrication of nanocrystalline diamond films using hot filament CVD, *Diamond Relat. Mater.*, **13**, pp. 6–13.

129. Jin, S., and Moustakas, T.D. (1994). Effect of nitrogen on the growth of diamond films, *Appl. Phys. Lett.*, **65**, pp. 403–405.

130. Mort, J., Machonkin, M.A., and Okumura, K. (1991). Compensation effects in nitrogen-doped diamond thin films, *Appl. Phys. Lett.*, **59**, pp. 3148–3150.

131. Müller-Sebert, W., Wörner, E., Fuchs, F., Wild, C., and Koidl, P. (1996). Nitrogen induced increase of growth rate in chemical vapor deposition of diamond, *Appl. Phys. Lett.*, **68**, pp. 759–760.

132. Mankelevich, Y.A., Suetin, N.V., Smith, J.A., and Ashfold, M.N.R. (2002). Investigations of the gas phase chemistry in a hot filament CVD reactor operating with $CH_4/N_2/H_2/$ and $CH_4/NH_3/H_2$ gas mixtures, *Diamond Relat. Mater.*, **11**, pp. 567–572.

133. Smith, J.A., Wills, J.B., Moores, H.S., Orr-Ewing, A.J., Ashfold, M.N.R., Mankelevich, Y.A., and Suetin, N.V. (2002). Effects of NH_3 and N_2 additions to hot filament activated CH_4/H_2 gas mixtures, *J. Appl. Phys.*, **92**(2), pp. 672–681.

134. Hirmke, J., Rosiwal, S.M., and Singer, R.F. (2008). Monitoring oxygen species in diamond hot-filament CVD by zircon dioxide sensors, *Vacuum*, **82**, pp. 599–607.

135. Corvin, R.B., Harrison, J.G., Catledge, S.A., and Vohra, Y.K. (2002). Gas-phase thermodynamic models of nitrogen-induced nanocrystallinity in chemical vapor-deposited diamond, *Appl. Phys. Lett.*, **80**, pp. 2550–2552.

Chapter 3

Advanced Carbon Aerogels for Energy Applications

Juergen Biener, Michael Stadermann, Matthew Suss, Marcus A. Worsley, Monika M. Biener, and Theodore F. Baumann
Nanoscale Synthesis and Characterization Laboratory, Lawrence Livermore National Laboratory, 7000 East Avenue, Livermore, CA 94550, USA
biener2@llnl.gov and baumann2@llnl.gov

Carbon aerogels are a unique class of high-surface-area materials derived by sol-gel chemistry. Their properties, such as high mass-specific surface area, electrical conductivity, environmental compatibility, and chemical inertness, make them very promising materials for many energy-related applications, specifically in view of recent developments in controlling their morphology and functionality. In this chapter, we will review the synthesis of monolithic resorcinol–formaldehyde-based carbon aerogels with hierarchical porosities for energy applications, including carbon nanotube and graphene composite carbon aerogels, as well as their functionalization by surface engineering. Applications that we will discuss include hydrogen and electrical energy storage, desalination, and catalysis.

Carbon-based Nanomaterials and Hybrids: Synthesis, Properties, and Commercial Applications
Edited by Hans-Jörg Fecht, Kai Brühne, and Peter Gluche

ISBN 978-981-4316-85-9 (Hardcover), 978-981-4411-41-7 (eBook)
www.panstanford.com

3.1 Introduction

The expression "the new carbon age" has been coined to express the rapid development of carbon-based nanomaterials that hold great technological promise for a variety of applications. Well-known examples are zero-dimensional bucky balls, one-dimensional carbon nanotubes (CNTs), and the relatively new class of two-dimensional graphene nanosheets, the subject of the 2010 Nobel Prize in Physics. Three-dimensional (3D) nanoporous carbons are another, equally important class of carbon nanomaterials that hold great technological promise for a variety of sustainable energy applications, including energy storage, adsorption, and catalysis [1–7]. All these new carbon nanomaterials have in common that they are built from sp^2-hybridized "graphitic" carbon atoms, thus providing them with mechanical strength and unique electronic properties. The utility of porous carbons is derived from their high surface area, 3D structure, chemical stability and abundance of carbon, and low mass density [8]. Among these materials, carbon aerogels (CAs) are specifically promising in that they possess a *tunable 3D hierarchical* morphology with ultrafine cell sizes and an electrically conductive framework (Fig. 3.1) that can be easily further modified by incorporation of materials such as CNTs and graphene and that are available as macroscopic, centimeter-sized monolithic materials.

Aerogels in general are a special class of open-cell foams that exhibit many fascinating properties, such as low mass densities, continuous porosities, and high surface areas. These unique properties are derived from the aerogel microstructure, which typically consists of 3D networks of interconnected nanometer-sized primary particles. Aerogels are prepared by sol-gel chemistry, a process that involves the formation and cross-linking of colloidal nanoparticles from molecular precursors. To preserve the tenuous solid network of the resulting inorganic or organic gels, one needs to use special drying techniques (i.e., supercritical drying, freeze-drying, etc.). Although the first aerogels based on silica gels were discovered already in 1931 [9], it took another 60 years until polymer-based CAs were developed by Pekala et al. [10, 11] in the late eighties at the Lawrence Livermore National Laboratory (LLNL). However, their technological potential, for example, as electrode materials, was immediately recognized [12]. Today, CAs have many promising applications in the field of supercapacitors and rechargeable batteries, catalysis, adsorbents, and thermal insulation [13–21].

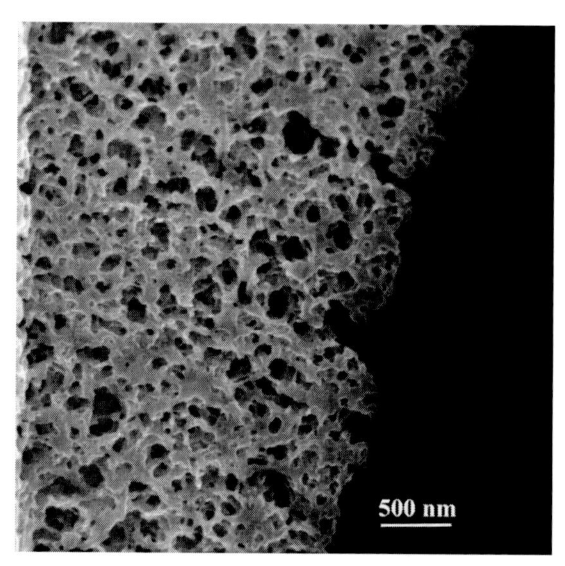

Figure 3.1 Characteristic three-dimensional hierarchical morphology of resorcinol-formaldehyde-based carbon aerogels (density of ~500 mg/cc).

CAs are prepared through simple sol-gel polymerization of organic precursors, such as resorcinol and formaldehyde, in aqueous solution and thus are mass producible. The sol-gel process yields highly cross-linked organic gels that then are supercritically dried and subsequently pyrolyzed in an inert atmosphere [10]. The pyrolysis transforms the organic aerogel precursor into a porous carbon network comprised of both amorphous and graphitic (turbostratic microcrystalline) regions. The graphitic domains within CAs, however, are typically quite small and contain a significant amount of disorder. Unlike many other porous carbon materials, CAs can be fabricated in a variety of forms, including monoliths and thin films, a feature that can be advantageous for many applications. Furthermore, the synthesis of CAs can be easily adjusted to create very different 3D architectures (see Fig. 3.1), which can be used to tailor the mass transport properties of CAs. This chapter first summarizes some of the most promising applications of CAs in the field of energy applications, including hydrogen and electrical energy storage, desalination, and catalysis, followed by an overview of recent progress in synthetic methodologies and the resulting morphological and functional diversity.

3.2 Applications

The most promising, emerging energy-related applications of CAs are hydrogen storage, electrical energy storage using electric double-layer capacitor (EDLC) technology, desalination using the capacitive deionization (CDI) technique, and their use as catalytic supports in, for example, fuel cells. All of these applications benefit from the tunability of the CA structure that provides a tool for improving device performance. In the following sections, we will first introduce these applications with an emphasis on functional requirements and then describe various concepts to implement the required functionality into the CA design.

3.2.1 Hydrogen Storage

The use of porous carbons for hydrogen storage has received significant attention [22–26]. Safe and efficient storage of hydrogen is considered one of the main challenges associated with utilization of hydrogen in the transportation sector [27]. An efficient hydrogen storage medium needs to be able to store a significant fraction of its own weight in hydrogen, preferable at noncryogenic temperatures, combined with fast loading/unloading kinetics. Porous carbons are promising due to their lightweight frameworks and their high accessible surface area, but their low hydrogen binding energies (~6 kJ/mol H_2) require cryogenic temperatures (77 K). The advantage of CAs over more traditional porous carbons is the flexibility that is associated with their synthesis. The goal is to design materials that have ultrahigh surface areas (up to 3,000 m^2/g) combined with pores that are slit shaped and have the optimum diameter for hydrogen physisorption (0.7–1 nm) [28, 29]. Strategies to increase the hydrogen-binding energy, and thus the operating temperature, include surface functionalization, for example, by metal doping or substitutional doping of carbon with boron or other light elements. The flexibility associated with CA synthesis allows for the incorporation of such modifiers into the carbon framework.

Finally, CAs can also be used as scaffolds to improve the stability and loading–unloading kinetics of other solid-state hydrogen storage materials, such as complex hydrides and borohydrides (BH_4^-) that offer high gravimetric and volumetric hydrogen capacities [18, 19]. Practical application of the scaffold approach requires the synthesis of mechanically robust CAs that have small-enough pores to physically confine the nanostructured hydride in combination with a large, accessible pore volume to minimize the gravimetric and volumetric capacity penalties associated with the use of the scaffold in a storage system. These structural refinements present a challenging trade-off in terms of porosity and mechanical properties. Maintaining small pore sizes while increasing the pore volume requires reducing the pore wall thickness, which, in turn, determines the mechanical integrity of the material.

3.2.2 Supercapacitors and Batteries

One of the most promising applications of CAs is use as electrode materials in EDLCs [30]. In these devices, electrical energy is stored in the form of ions accumulated on the surface of the material (see Fig. 3.2b,c), thus creating an intermediate between a battery and an electrostatic capacitor [31]. EDLCs are an ideal compliment to batteries in devices with peak power demands above the base level, where they extend the life of the battery. A typical example is the use of supercapacitors for regenerative braking in electric and hybrid cars. However, the high cost of EDLCs, as well as their limited energy density, has so far prevented them from replacing batteries more widely [32]. CAs are ideally suited as electrode materials as they are chemically inert, highly conductive, and environmentally friendly, can be made from cheap and abundant raw materials, and, most importantly, provide high specific surface areas combined with a fully tunable 3D structure. Specifically the tunability of the CA structure provides a tool for improving energy and power densities, as both depend on the pore structure of the electrode. As in the case of hydrogen storage, surface functionalization can be used to further increase the storage capacity.

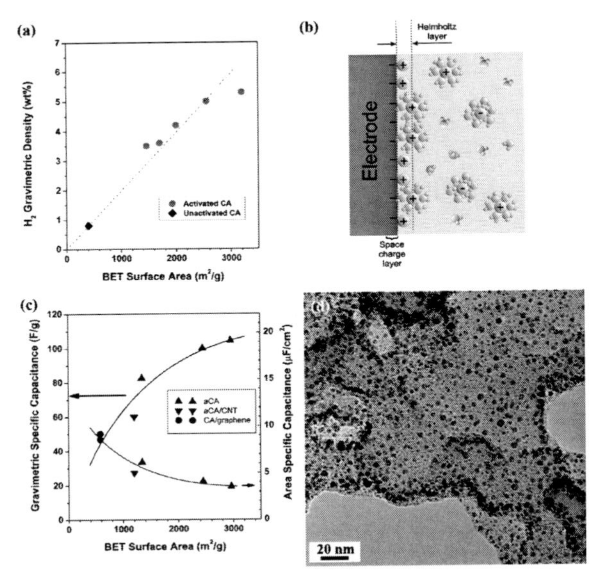

Figure 3.2 The most promising emerging energy-related applications of CAs are hydrogen and electrical energy storage, desalination, and their use as catalytic supports. (a) Hydrogen uptake at 77 K in activated CAs scales linearly with the BET surface area up to 2,500 m^2/g, yielding gravimetric densities up to 5 wt% hydrogen and volumetric capacities up to 29 g hydrogen/L. CAs can also store electrical energy using the EDL phenomena, which relies on the accumulation of ions on the surface of the material. The EDL phenomena can also be used for desalination using capacitive deionization technology. (b) The EDL model showing the accumulation of cations at a negatively polarized electrode surface in contact with an electrolyte. In the case of an ideally polarizable electrode, no charge transfer takes place across the electrode/electrolyte interface, leading to a capacitor-like separation of positive and negative charges. The atomic-scale dimensions of the EDL results in much higher capacities compared to that of a conventional parallel-plate capacitor. (c) Gravimetric- and area-specific capacitance versus surface area for various activated CAs (▲) as well as activated CA/CNT (▼) and CA/graphene (•) composite materials. The gravimetric-specific capacitance increases, and the area-specific capacitance decreases with increasing surface area. (d) Micrograph from a platinum-loaded CA prepared by ALD. The material exhibits high catalytic activity for the oxidation of CO, even at loading levels as low as ~0.05 mg Pt/cm^2. *Abbreviations*: BET, Brunauer–Emmet–Teller; ALD, atomic-layer deposition.

3.2.3 Capacitive Deionization

CAs also have considerable potential to significantly improve desalination efficiency through a technique known as CDI. Like the supercapacitor, CDI relies on the formation of an EDL to store charge (Fig. 3.2b,c). In CDI, however, the goal of the charge storage is not energy storage but removal of ions from the electrolyte to create a clean water source. To date, CDI has been successfully applied to the desalination of lower-salinity (brackish) water streams that typically contain in the order of 1,000 ppm of total dissolved salts (TDS) (for comparison, seawater is about 35,000 ppm TDS) [33–35]. CDI systems developed in the 1960s to early 1990s for water desalination utilized porous carbon-based electrode materials such as activated carbons [36–39]. The first CDI system that used CA electrodes was developed in the 1990s by Farmer et al. at the LLNL [34, 40]. CAs offer several advantages over activated carbons. Specifically, the ability to tune the pore size distribution in hierarchically structured CAs can be used to improve the energy efficiency of a CDI system by reducing ionic transport losses, while maintaining a high capacitance.

3.2.4 Catalysis

Another promising energy application for CAs is their use as electrode materials and catalyst support in proton-exchange membrane (PEM) fuel cells. For this application, the electrode needs to combine high electrochemical conductivity, high surface area, and a pore structure that allows for good contact between the catalytically active supported metal nanoparticles, the polymer electrolyte, and the gas phase to minimize mass-transport losses [41]. The advantage of CAs over other more traditional carbon supports such as commercial carbon blacks is that their surface area, pore size, and pore volume can be tailored independently from each other [42]. They also offer superior electrical conductivity due to their 3D morphology, which reduces the electric losses in the electrode, and they are available as monolithic structures. The required catalytic activity needs to be introduced by metal loading of the otherwise inert CA, and an example of a platinum-loaded CA is shown in Fig. 3.2d.

3.3 Design of Carbon Aerogels

3.3.1 Synthesis

As previously stated, the synthesis of CAs begins with the sol-gel polymerization of organic precursors to form a highly cross-linked organic gel. While a variety of multifunctional monomers have been utilized for this type of reaction, the most commonly used precursors for the synthesis of CAs are resorcinol (1,3-dihydroxybenzene) and formaldehyde (Fig. 3.3) [43]. In this particular system, the resorcinol serves as the multifunctional monomer that contains three reactive sites at the 2, 4, and 6 positions on the benzene ring. The resorcinol reacts with the formaldehyde in aqueous solution, initiated by a catalyst (either acid or base), to form mixtures of addition and condensation products. The formation of the resorcinol–formaldehyde (RF) reaction products involves two main steps, (1) the addition of formaldehyde to resorcinol to form hydroxymethyl derivatives and (2) subsequent condensation of these hydroxymethyl derivatives to form methylene and methylene ether bridges between resorcinol molecules. The catalyst plays a critical role in activating the monomers for the addition and condensation reactions. For example, basic catalysts, such as carbonate or hydroxide, promote the formation of resorcinol anions that are more reactive toward the addition of formaldehyde than uncharged resorcinol molecules. Acid catalysts, by contrast, typically react with the formaldehyde to initiate polymerization [44]. These condensation reactions lead to the formation of nanometer-size clusters of the RF polymer in the reaction solution. These colloidal particles then aggregate and react with one another through surface hydroxymethyl functional groups to form a 3D gel structure that fills the original volume of the aqueous solution.

To obtain an aerogel, the solvent that occupies the pores of the gel must be removed in a way that preserves the tenuous network of the solid phase [45]. Direct evaporation of the solvent from the pores of the gel typically leads to collapse of the gel architecture due to capillary stress. Materials prepared by evaporative drying (referred to as xerogels) have significantly higher densities and possess less open porosity than the original gel architecture. One approach to retain the low-density structure of an organic gel upon drying is to exceed the critical point of the solvent within the pores.

The supercritical fluid has no meniscus and therefore does not generate damaging capillary forces during drying. As an example, supercritical carbon dioxide extraction is commonly used to dry organic RF gels [10]. In this process, the water in the pores of the gel is first exchanged with an appropriate organic solvent, such as acetone, since water is not miscible with liquid carbon dioxide. The gels are then placed in a pressure vessel, wherein liquid carbon dioxide is fully exchanged for the organic solvent. The gel is dried by taking the carbon dioxide above its critical point (T_c = 31°C, P_c = 7.38 MPa) and then slowly venting the supercritical CO_2 from the pressure vessel. Organic aerogels obtained by supercritical drying are highly porous (50–90%) and retain the original solid-phase architecture of interconnected polymer particles. In the case of RF aerogels, the dried materials are typically isolated as transparent red or orange monoliths.

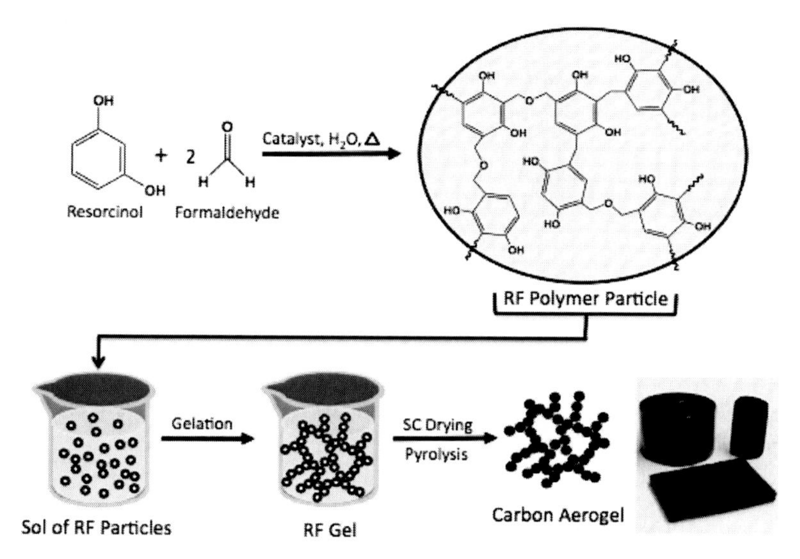

Figure 3.3 Scheme depicting the synthesis of CAs derived from resorcinol and formaldehyde.

To potentially lower the cost and time associated with aerogel processing, other drying approaches have been investigated for the fabrication of aerogel-like materials without the need for supercritical extraction. For example, freeze-drying has been used to dry organic RF gels without a significant collapse of microstructure [46, 47]. In this method, the solvent within the pores of the gel is frozen and subsequently removed through sublimation, thus avoiding capillary stresses. While this approach does result in slight densification of the network structure, the obtained materials (termed "cryogels") still exhibit a high degree of mesoporosity. Alternatively, evaporative drying using a nonpolar organic solvent with low surface tension, such as cyclohexane, has been used to produce low-density materials (ambigels) that retain ~80% of the porosity of aerogels [48, 49].

CAs are derived from the pyrolysis of the organic aerogel architecture in an inert atmosphere, such as nitrogen or argon. During pyrolysis of RF aerogels, significant mass losses are observed (~50%), accompanied by volumetric shrinkage (>50%) of the monolith parts. The mass loss and dimensional changes are associated with the decomposition and conversion of the network polymer to a carbon skeleton that consists of both amorphous and graphitic (turbostratic microcrystalline) regions [8]. Despite the mass loss and dimensional changes that occur during pyrolysis, these materials retain the highly porous architecture of the organic aerogel. The main structural differences observed in CAs are smaller feature sizes (particles and pores) relative to the starting organic aerogel and the evolution of microstructures (i.e., micropores, crystallites) in the carbon framework. The degree to which these transformations occur is related to the carbonization temperature. For RF aerogels, the majority of the mass loss occurs in the range of 300–800°C. At 1,000°C, the conversion of the RF aerogel to pure CAs is practically complete, as only small amounts of hydrogen and oxygen are detected in the pyrolyzed material. Most phenolic resins, however, do not completely graphitize, even when treated at temperatures above 2,500°C. CAs can be fabricated in a variety of different forms, including thin films and conformable monoliths (Fig. 3.3), and exhibit a range of unique bulk physical properties. For example, CAs derived from RF polymers typically possess high accessible surface areas (up to ~1,000 m^2/g) and large internal pore volumes. Additionally, these materials are electrically conductive following pyrolysis due to the continuous network of covalently bonded carbon particles [50].

The technological promise of CAs comes from the fact that their bulk physical properties can be controlled through the sol-gel reaction chemistry. Several factors of the polymerization reaction have a significant impact on network formation in these materials.

The choice of sol-gel precursors, polymerization catalyst, and reaction solvent can each be used to control the microstructure of the resultant gel. As an example, the amount and type of polymerization catalyst used in the sol-gel reaction influences the nucleation, growth, and interconnectivity of the primary particles that comprise the aerogel framework.

In the case of base-catalyzed RF aerogels, the size and number of polymer clusters formed during polymerization is controlled by the molar ratio of resorcinol to catalyst (R/C ratio) used in the reaction. Formulations that utilize high R/C ratios ($R/C \geq 500$) typically produce gels with larger primary particles as compared to those generated from low R/C ratios ($R/C \sim 50$) [51]. The morphology and spatial arrangement of these particles, in turn, determine the bulk physical properties of the CA. For instance, electrical conductivity in CAs occurs through the movement of charge carriers through individual carbon particles and "hopping" of these carriers between adjacent carbon particles [50]. Therefore, charge transport is highly dependent on both particle morphology and interconnectivity of the carbon network. Likewise, a number of other bulk properties, such as specific surface area, average pore size, compressive modulus, and thermal conductivity, correlate with the network architecture and, therefore, can be tuned through the reaction chemistry. Beyond the reaction chemistry, the bulk properties of these materials can be further modified either through surface treatment (i.e., activation, deposition) of the aerogel or through the direct incorporation of modifiers into the skeletal framework of the CA (Fig. 3.4). The following sections provide an overview of the different techniques that have been utilized to enhance particular attributes of these materials.

3.3.2 Activation

As prepared, CAs can have specific surface areas ranging from 200 m^2/g to 1,000 m^2/g, depending on the reaction formulation. The accessible surface areas in these materials can be further increased through chemical or thermal activation processes. Chemical activation can be performed through heating the carbon (or carbon precursor) in concentrated acids (H_3PO_4) or bases (i.e., KOH) [52, 53]. Thermal activation of CAs involves the controlled burn-off of carbon from the network structure in an oxidizing atmosphere, such

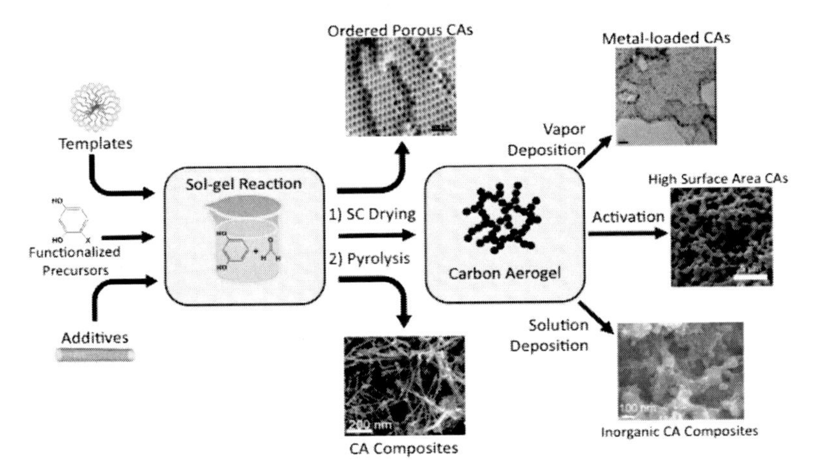

Figure 3.4 Scheme showing the versatility of CA synthesis. The architecture of CAs can be modified either during the sol-gel polymerization step, through the introduction of additives or templates to the reaction mixture, or through gas- or solution-phase reactions on the surfaces of the CA framework after the pyrolysis step.

as steam, air, or carbon dioxide, at elevated temperatures [54–59]. In both cases, activation creates new micropores and opens closed porosity in the CA framework, leading to an increase in the overall surface area. While a variety of different activation techniques have been used with CAs, thermal activation with carbon dioxide has proved to be the most effective method for generating high-surface-area CAs with uniform micropore size distributions. Carbon dioxide reacts with carbon from the CA network according to the overall reaction

$$C + CO_2 \leftrightarrow 2CO, \quad (3.1)$$

which is also known as the Boudouard equilibrium. With this approach, the BET surface area increases linearly with carbon burn-off at constant temperature.

Access to ultrahigh surface areas in activated CAs also requires careful design of the preactivated carbon framework, as the morphology of the network structure will ultimately determine the textural properties of the activated material. The microstructure of traditional RF-derived CAs, consisting of nanometer-size carbon particles and tortuous pore structures, can both limit the surface areas attainable through activation and lead to inhomogeneous burn-off in monolithic samples. Thermal activation of CAs with larger structural features (micron-size pores and ligaments) was recently shown to afford mechanically robust bulk materials with BET surface areas in excess of 3,000 m^2/g (Fig. 3.5) [54]. These values are greater than the surface area of a single graphene sheet (2,630 m^2/g, if both graphene surfaces are taken into account). Presumably, edge termination sites constitute a substantial fraction of the surface area in these activated CAs, as is the case for traditional high-surface-area activated carbons. As shown in Fig. 3.5, increasing activation does not only increase the surface area of the CA but also shifts the pore size distribution to larger values. Activated CAs have been investigated for a variety of applications, including catalysis, gas storage, and EDLCs.

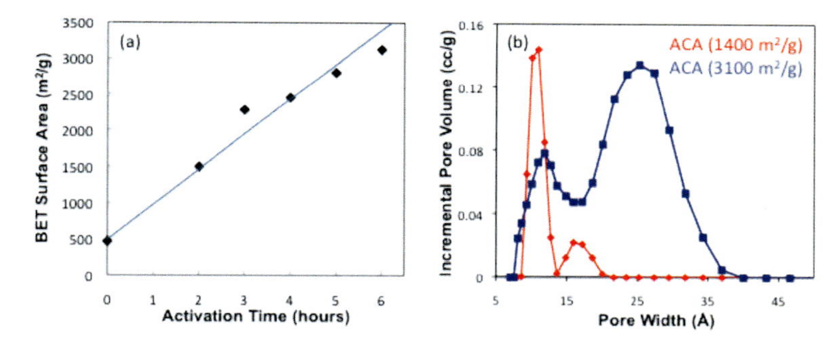

Figure 3.5 (a) BET surface area of ACAs as a function of activation time using CO_2 at 950°C. (b) Micropore size distribution for two ACAs with different BET surface areas. *Abbreviation*: ACA, activated carbon aerogel.

3.3.3 Incorporation of Modifiers

To further optimize the properties of CAs for specific applications, recent efforts have focused on modification of the CA framework through the incorporation of additives, such as catalyst nanoparticles or CNTs, directly into the carbon framework. The flexibility of CA synthesis readily allows for systematic modification of the extended network structure. For example, functionalized sol-gel precursors (i.e., resorcinol derivatives) can be used in the polymerization reaction to introduce specific functional groups into the organic and CA framework [60–64]. This approach has been used to homogeneously incorporate metal nanoparticles into CA matrices (Fig. 3.6).

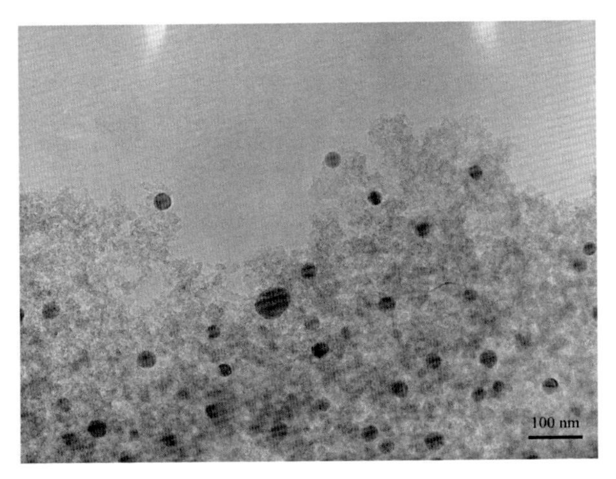

Figure 3.6 TEM image of a copper-doped CA that was prepared through the sol-gel polymerization of 2,4-dihydroxybenzoic acid (a resorcinol derivative) with formaldehyde. After gel formation, metal ions were uniformly incorporated into the organic polymer gel through the carboxylic acid groups of the resorcinol derivative. The metal ions are then reduced to metal nanoparticles (dark spots) during the carbonization step. *Abbreviation*: TEM, transmission electron microscopy.

Another approach to modify the CA framework involves the integration of additives, such as CNTs or graphene sheets, into the sol-gel reaction [65–72]. Sol-gel polymerization of the RF precursors in the presence of the additives leads to the formation of composite structures in which the additives become part of the primary carbon network structure (Fig. 3.7). For example, CNT–CA nanocomposites have been prepared through the sol-gel polymerization of resorcinol and formaldehyde in an aqueous surfactant-stabilized suspension of double-walled CNTs. The composite gels were then supercritically dried and subsequently carbonized to yield monolithic CA structures containing uniform dispersions of CNTs. These composite materials exhibit enhanced electrical, thermal, and mechanical properties relative to pristine CAs of equivalent density [68, 69]. A similar methodology has also been used in the fabrication of ultralow-density foams comprised of either CNTs [67] or individual graphene sheets [65, 66]. In this case, the RF-derived carbon is used as an electrically conductive binder to cross-link the single-walled CNTs or graphene sheets that define the foam architecture (Fig. 3.3). This approach produces ultralow-density monolithic solids (as low as ~10 mg cm^{-3}) that simultaneously exhibit remarkable mechanical stiffness, very large elastic strains, and high electrical conductivity. This process also provides the versatility to generate these monolithic foams in conformable shapes for different applications.

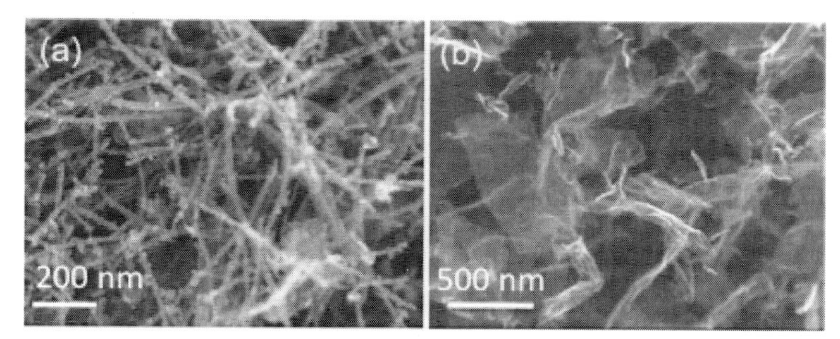

Figure 3.7 SEM images of carbon foams comprising of (a) randomly oriented single-walled CNT bundles and (b) individual graphene sheets that are held together using an RF-derived CA as an electrically conductive binder. *Abbreviation*: SEM, scanning electron microscopy.

3.3.4 Templating

Templates can also be introduced into the sol-gel polymerization reaction to impart specific structural elements to the porous solid under construction, for example, to improve the mass-transport properties of CAs [73–79]. After gel formation, these templates can be removed, either chemically or thermally, to yield defined features within the aerogel framework. For example, colloidal crystals (close-packed arrays of monodisperse spheres) have been used as templates for the preparation of three-dimensionally ordered macroporous CAs (Fig. 3.8) [77]. In this approach, the interstitial regions of the colloidal crystal are infused with the RF reaction solution to form a composite gel structure. After removal of the template, the replicate gel can be dried and carbonized to yield a macroporous CA containing periodic voids, the diameter of which is controlled by the size of the template spheres. Using this technique, materials with bimodal pore structures can be prepared in which the large ordered cavities (d = 0.1–1 μm) formed by the template are continuous and interconnected, while smaller meso- and microporous channels run continuously throughout the aerogel wall. Hierarchically porous structures of this type present a number of advantages over unimodal carbon structures in terms of diffusion efficiency and surface area and thus have utility as catalyst supports or electrodes for electrochemical devices.

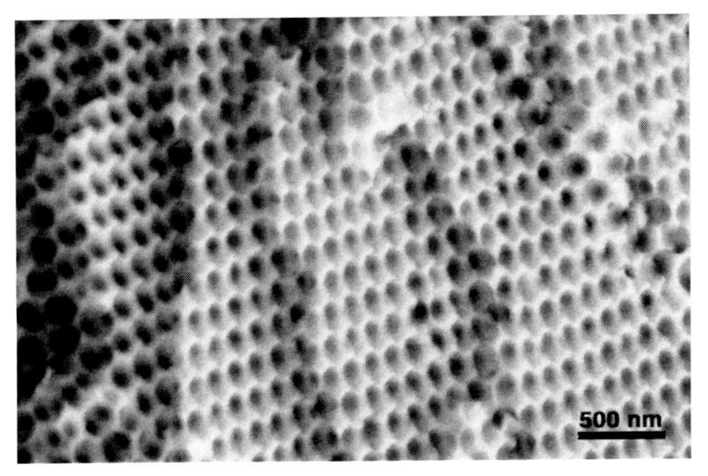

Figure 3.8 SEM image of an ordered macroporous CA prepared using a colloidal crystal of polystyrene spheres as a template.

3.3.5 Surface Functionalization

Surface functionalization is an important method to improve the performance of CAs. For example, the charge storage capacity of a CA supercapacitor can be vastly increased by coating the surfaces with reducible transition metal oxides, thus creating a pseudocapacitor. Using a CA as a catalyst (e.g., in fuel cell applications) also requires the incorporation of metal nanoparticles into the CA structure to add catalytic activity to the otherwise inert CA (Fig. 3.9). Although loading of CAs with functional materials seems to be simple in theory, it is difficult in practice due to the high aspect ratio of CAs. For example, a centimeter-size CA sample with 100 nm pores has an aspect ratio of >10^5. High utilization of the incorporated materials, which is especially an issue for expensive materials such as the catalytically active noble metals (platinum and platinum alloys), requires that the functional material be highly dispersed to maximize its surface area. Furthermore, the functional material needs to be resistant against sintering and stably anchored to the carbon support. Solution impregnation using a suitable metal salt, either during the synthesis of the polymer gel [60, 61, 64] or at a later stage by impregnation of the pyrolized CA [42, 63, 80], has traditionally been used to synthesize metal-loaded CAs. However, both approaches have their limitations: the former can interfere with the sol-gel process or can lead to poisoning of the catalyst, for example, by coating Pt nanoparticles with a passivating carbon layer [63].

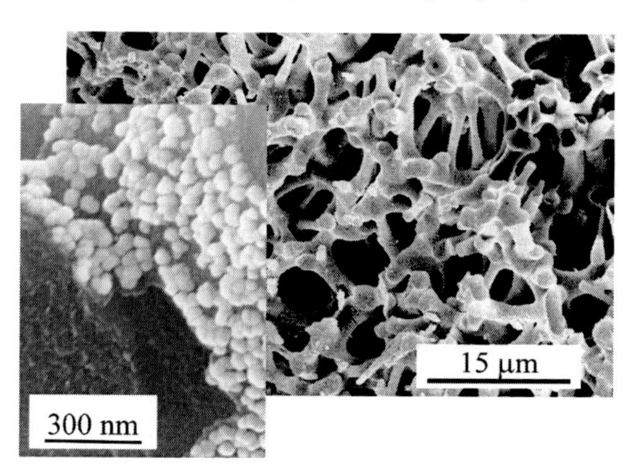

Figure 3.9 SEM micrographs of a Ru-loaded CA at different magnification levels. The Ru nanoparticles were deposited using bis(cyclopentadienyl)-ruthenium/air ALD processes.

Despite these difficulties, there are many examples where the surfaces of the aerogel framework have been modified after the carbonization step through solution-phase reactions within the open pore volume of the aerogel. Typically, these structures are sufficiently robust to withstand the compressive forces associated with rewetting and drying of the material. For example, self-limiting electroless deposition processes have been used to "paint" the interior structure of CA materials with electroactive species, such as MnO_2, for the design of new capacitors [81–84]. Inorganic CA composites have also been prepared through the sol-gel polymerization of inorganic species on the internal surfaces of monolithic CA parts [85–87].

With this approach, the inorganic species (metal oxides such as SiO_2, TiO_2, or ZnO) form a conformal overlayer on the primary ligament structure of the CA. Nanocomposites formed from carbon and these metal oxides have the potential to exhibit enhanced functional properties for catalysis and energy storage applications. This approach has also been used to access new surface compositions through the carbothermal reduction of the oxide overlayer to metal carbides or carbonitrides. For example, recent work has shown that the SiO_2 coating on the ligaments of an activated CA can be converted to SiC by heating the composite material to 1,400°C in an argon atmosphere (Fig. 3.10). These carbide- and carbonitride-coated CA materials exhibit improved thermal stability in oxidizing environments.

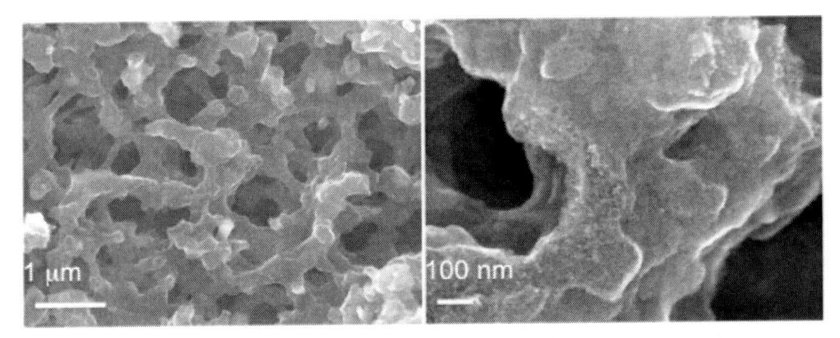

Figure 3.10 SEM images of a SiC-coated CA at different magnifications prepared through the carbothermal reduction of a SiO_2-coated CA at 1,400°C under Ar.

An alternative to the solution impregnation discussed above is vapor-phase deposition techniques. However, vacuum thin-film techniques that, for example, are commonly used to deposit thin catalyst layers directly onto the membrane or the gas diffusion layer of PEM fuel cells [88] cannot take full advantage of the high aspect ratio/mesoporosity of CAs as they are only able to coat the outer surface. Nevertheless, magnetron sputtering has been used to deposit Pt on thin CA sheets [89, 90]. A similar restriction applies to chemical vapor deposition (CVD) techniques that can be successfully applied to macrocellular foams [91] but typically result in inhomogeneous or incomplete coatings on micromesoporous substrates.

ALD, on the other hand, is a special variant of the CVD technique that uses a suitable pair of sequential, self-limiting surface reactions

(Fig. 3.11) [92] and is thus suited like no other method to deposit material onto the internal surfaces of ultra-high-aspect-ratio materials such as CAs. The technique does not affect the morphology of the aerogel template and offers excellent control over the loading level by simply adjusting the number of ALD cycles. Continuous layers as well as nanoparticles can be grown depending on the actual surface chemistry. Both oxidic and metallic films and nanoparticles can be deposited, and the technique has been successfully employed to deposit a variety of materials, including W [93], Ru [94], Pt [17], Cu [95], TiO_2 [96], and ZnO [97], on various aerogel templates. The two most important metals for the fuel cell application, Pt and Ru, can be deposited following the ALD recipes developed by Aalton et al. [98, 99], and a typical example is shown in Fig. 3.9. As both processes run in the same temperature window they also can be combined to deposit Pt–Ru alloy nanoparticles. The technique has already been successfully used to deposit Pt nanoparticles on PEM fuels cell electrodes [100].

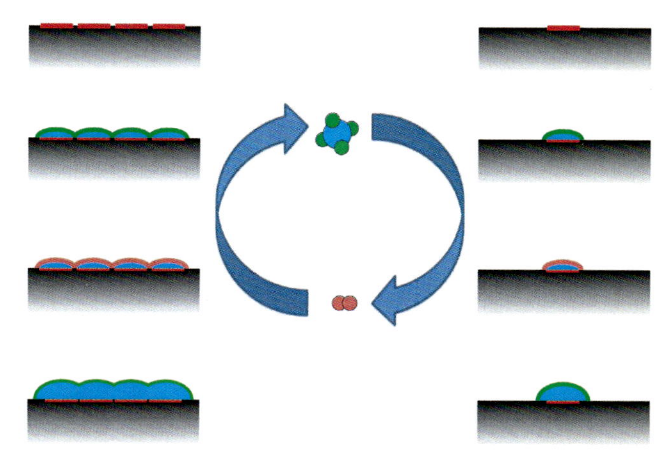

Figure 3.11 ALD employs sequential, self-limiting surface reactions to overcome diffusion limitations in high-aspect-ratio materials. The underlying principle is that the adsorption of the metal containing precursor poisons the surface, thus preventing further precursor uptake (green surface termination), and that the reactive surface state can be regenerated by exposure to the second reactant (red in this example). Both conformal films (left) and individual nanoparticles (right) can be grown depending on the initial nucleation density, which can be influenced by surface treatments.

More recently, surface modification using vapor-phase techniques has also been extended to the use of CAs as substrates for the direct growth of CNTs by CVD [101]. By engineering the pore structure of the CA substrate, uniform CNT yield can be achieved throughout the free internal pore volume of CA monoliths with macroscopic dimensions.

3.4 Summary

CAs have become a very promising material for many energy-related applications due to recent progress in the synthesis of mechanically robust, monolithic CAs with tunable hierarchical morphologies, the incorporation of modifiers such as CNTs and graphene, and the development of suitable surface funtionalization techniques such as ALD. The most promising energy-related applications of CAs are hydrogen and electrical energy storage, desalination, and catalysis. Beyond being functional materials by themselves, CAs have also become very promising scaffold materials due to their mechanical and chemical stability that can be used to stabilize other nanomaterials by confinement. These CA-stabilized nanomaterials combine the faster kinetics of nanomaterials with the stability of CAs. Further performance improvements, for example, in storage capacity and loading/unloading dynamics will require a better understanding of the correlations between morphology and surface chemistry and properties such as hydrogen and electrical energy storage capacity, as well as the effect of crystallinity on the electronic and mechanical properties of CAs. Increasing the storage density will require the development of new high-density, high-surface-area CAs.

Acknowledgments

Work at the LLNL was performed under the auspices of the US Department of Energy (DOE) by the LLNL under Contract DE-AC52-07NA27344 and funded in part by the DOE Office of Energy Efficiency and Renewable Energy. The desalination work was funded in part by the State of California's Proposition 50; funds are administered by the Department of Water Resources.

References

1. Stein, A., Wang, Z., and Fierke, M.A. (2009). Functionalization of porous carbon materials with designed pore architecture, *Adv. Mater.*, **21**, pp. 265–293.
2. Hu, Y.S., Adelhelm, P., Smarsly, B., Hore, S., Antonietti, M., and Maier, J. (2007). Synthesis of hierarchically porous carbon monoliths with highly ordered microstructure and their application in rechargeable lithium batteries with high-rate capability, *Adv. Funct. Mater.*, **17**, pp. 1873–1878.
3. Chmiola, J., Yushin, G., Gogotsi, Y., Portet, C., Simon, P., and Taberna, P.L. (2006). Anomalous increase in carbon capacitance at pore sizes less than 1 nanometer, *Science*, **313**, pp. 1760–1763.
4. Tabata, S., Iida, H., Horie, T., and Yamada, S. (2010). Hierarchical porous carbon from cell assemblies of rice husk for in vivo applications, *Med. Chem. Commun.*, **1**, pp. 136–138.
5. Lee, K., Lytle, J., Ergang, N., Oh, S., and Stein, A. (2005). Synthesis and rate performance of monolithic macroporous carbon electrodes for lithium-ion secondary batteries, *Adv. Funct. Mater.*, **15**, pp. 547–556.
6. Chai, G., Shin, I., and Yu, J.S. (2004). Synthesis of ordered, uniform, macroporous carbons with mesoporous walls templated by aggregates of polystyrene spheres and silica particles for use as catalyst supports in direct methanol fuel cells, *Adv. Mater.*, **16**, pp. 2057–2061.
7. Biener, J., Stadermann, M., Suss, M., Worsley, M.A., Biener, M.M., Rose, K.A., and Baumann, T.F. (2011). Advanced carbon aerogels for energy applications, *Energy Environ. Sci.*, **4**, pp. 656–667.
8. Fricke, J., and Petricevic, R. (2002). "Carbon aerogels," in *Handbook of Porous Solids*, eds. Schüth, F., Sing, K.S.W., and Weitkamp, J. (Wiley-VCH, Weinheim).
9. Kistler, S.S. (1931). Coherent expanded aerogels and jellies, *Nature*, **127**, pp. 741–741.
10. Pekala, R.W. (1989). Organic aerogels from the polycondensation of resorcinol with formaldehyde, *J. Mater. Sci.*, **24**, pp. 3221–3227.
11. Pekala, R.W. (1989). Low density, resorcinol-formaldehyde aerogels, *US Patent* 4873218.
12. Mayer, S.T., Pekala, R.W., and Kaschmitter, J.L. (1993). The aerocapacitor—an electrochemical double-layer energy-storage device, *J. Electrochem. Soc.*, **140**, pp. 446–451.

13. Nielsen, T.K., Boì̂senberg, U., Gosalawit, R., Dornheim, M., Cerenius, Y., Besenbacher, F., and Jensen, T.R. (2009). A reversible nanoconfined chemical reaction, *ACS Nano*, **4**, pp. 3903–3908.
14. Li, G.-R., Feng, Z.-P., Ou, Y.-N., Wu, D., Fu, R., and Tong, Y.-X. (2010). Mesoporous MnO_2/carbon aerogel composites as promising electrode materials for high-performance supercapacitors, *Langmuir*, **26**, pp. 2209–2213.
15. Rolison, D.R., Long, J.W., Lytle, J.C., Fischer, A.E., Rhodes, C.P., McEvoy, T.M., Bourg, M.E., and Lubers, A.M. (2009). Multifunctional 3D nanoarchitectures for energy storage and conversion, *Chem. Soc. Rev.*, **38**, pp. 226–252.
16. Tian, H.Y., Buckley, C.E., Wang, S.B., and Zhou, M.F. (2009). Enhanced hydrogen storage capacity in carbon aerogels treated with KOH, *Carbon*, **47**, pp. 2128–2130.
17. King, J.S., Wittstock, A., Biener, J., Kucheyev, S.O., Wang, Y.M., Baumann, T.F., Giri, S.K., Hamza, A.V., Bäumer, M., and Bent, S.F. (2008). Ultralow loading Pt nanocatalysts prepared by atomic layer deposition on carbon aerogels, *Nano Lett.*, **8**, pp. 2405–2409.
18. Gross, A.F., Vajo, J.J., Van Atta, S.L., and Olson, G.L. (2008). Enhanced hydrogen storage kinetics of $LiBH_4$ in nanoporous carbon scaffolds, *J. Phys. Chem. C*, **112**, pp. 5651–5657.
19. Feaver, A., Sepehri, S., Shamberger, P., Stowe, A., Autrey, T., and Cao, G. (2007). Coherent carbon cryogel-ammonia borane nanocomposites for H_2 storage, *J. Phys. Chem. B*, **111**, pp. 7469–7472.
20. Du, H.D., Li, B.H., Kang, F.Y., Fu, R.W., and Zeng, Y.Q. (2007). Carbon aerogel supported Pt-Ru Catalysts for using as the anode of direct methanol fuel cells, *Carbon*, **45**, pp. 429–435.
21. Kabbour, H., Baumann, T.F., Satcher, J.H., Saulnier, A., and Ahn, C.C. (2006). Toward new candidates for hydrogen storage: high-surface-area carbon aerogels, *Chem. Mater.*, **18**, pp. 6085–6087.
22. McNicholas, T.P., Wang, A.M., O'Neill, K., Anderson, R.J., Stadie, N.P., Kleinhammes, A., Parilla, P., Simpson, L., Ahn, C.C., Wang, Y.Q., Wu, Y., and Liu, J. (2010). H_2 storage in microporous carbons from peek precursors, *J. Phys. Chem. C*, **114**, pp. 13902–13908.
23. Guan, C., Wang, K., Yang, C., and Zhao, X.S. (2009). Characterization of a zeolite-templated carbon for H_2 storage application, *Micropor. Mesopor. Mater.*, **118**, pp. 503–507.
24. Xia, K.S., Gao, Q.M., Wu, C.D., Song, S.Q., and Ruan, M.L. (2007). Activation, characterization and hydrogen storage properties of the mesoporous carbon CMK-3, *Carbon*, **45**, pp. 1989–1996.

25. Yang, Z.X., Xia, Y.D., and Mokaya, R. (2007). Enhanced hydrogen storage capacity of high surface area zeolite-like carbon materials, *J. Am. Chem. Soc.*, **129**, pp. 1673–1679.

26. Panella, B., Hirscher, M., and Roth, S. (2005). Hydrogen adsorption in different carbon nanostructures, *Carbon*, **43**, pp. 2209–2214.

27. Schlapbach, L., and Züttel, A. (2001). Hydrogen-storage materials for mobile applications, *Nature*, **414**, pp. 353–358.

28. Patchkovskii, S., Tse, J.S., Yurchenko, S.N., Zhechkov, L., Heine, T., and Seifert, G. (2005). Graphene nanostructures as tunable storage media for molecular hydrogen, *Proc. Natl. Acad. Sci.*, **102**, pp. 10439–10444.

29. Murata, K., Kaneko, K., Kanoh, H., Kasuya, D., Takahashi, K., Kokai, F., Yudasaka, M., and Iijima, S. (2002). Adsorption mechanism of supercritical hydrogen in internal and interstitial nanospaces of single-wall carbon nanohorn assembly, *J. Phys. Chem. B*, **106**, pp. 11132–11138.

30. Pekala, R.W., Farmer, J.C., Alviso, C.T., Tran, T.D., Mayer, S.T., Miller, J.M., and Dunn, B. (1998). Carbon aerogels for electrochemical applications, *J. Non-Cryst. Solids*, **225**, pp. 74–80.

31. Conway, B.E. (1991). Transition from "supercapacitor" to "battery" behavior in electrochemical energy storage, *J. Electrochem. Soc.*, **138**, pp. 1539–1548.

32. "Technology and applied R&D needs for electrical energy storage," in BES Workshop (2007).

33. Gabelich, C.J., Tran, T.D., and Suffet, I.H. (2002). Electrosorption of inorganic salts from aqueous solution using carbon aerogels, *Environ. Sci. Technol.*, **36**, pp. 3010–3019.

34. Farmer, J.C., Fix, D.V., Mack, G.V., Pekala, R.W., and Poco, J.F. (1996). Capacitive deionization of NaCl and $NaNO_3$ solutions with carbon aerogel electrodes, *J. Electrochem. Soc.*, **143**, pp. 159–169.

35. Biesheuvel, P.M., and van der Wal, A. (2010). Membrane capacitive deionization, *J. Membr. Sci.*, **346**, pp. 256–262.

36. Oren, Y. (2008). Capacitive deionization (CDI) for desalination and water treatment—past, present and future (a review). *Desalination*, **228**, pp. 10–29.

37. Murphy, G.W., and Tucker, J.H. (1966). The demineralization behavior of carbon and chemically-modified carbon electrodes, *Desalination*, **1**, pp. 247–259.

38. Johnson, A.M., and Newman, J. (1971). Desalting by means of porous carbon electrodes, *J. Electrochem. Soc.*, **118**, pp. 510–517.

39. Oren, Y., and Soffer, A. (1982). Water desalting by means of electrochemical parametric pumping. 1. The equilibrium properties of a batch unit cell, *J. Appl. Electrochem.*, **13**, pp. 473–487.

40. Farmer, J.C., Fix, D.V., Mack, G.V., Pekala, R.W., and Poco, J.F. (1996). Capacitive deionization of NH_4ClO_4 solutions with carbon aerogel electrodes, *J. Appl. Electrochem.*, **26**, pp. 1007–1018.

41. Dicks, A.L. (2006). The role of carbon in fuel cells, *J. Power Sources*, **156**, pp. 128–141.

42. Marie, J., Chenitz, R., Chatenet, M., Berthon-Fabry, S., Cornet, N., and Achard, P. (2009). Highly porous PEM Fuel cell cathodes based on low density carbon aerogels as Pt-support: experimental study of the mass-transport losses, *J. Power Sources*, **190**, pp. 423–434.

43. Al-Muhtaseb, S.A., and Ritter, J.A. (2003). Preparation and properties of resorcinol-formaldehyde organic and carbon gels, *Adv. Mater.*, **15**, pp. 101–114.

44. Brandt, R., Petricevic, R., Pröbstle, H., and Fricke, J. (2003). Acetic acid catalyzed carbon aerogels, *J. Porous Mater.*, **10**, pp. 171–178.

45. Heinrich, T., Klett, U., and Fricke, J. (1995). Aerogels—nanoporous materials part I: sol-gel process and drying of gels, *J. Porous Mater.*, **1**, pp. 7–17.

46. Yamamoto, T., Sugimoto, T., Suzuki, T., Mukai, S.R., and Tamon, H. (2002). Preparation and characterization of carbon cryogel microspheres, *Carbon*, **40**, pp. 1345–1351.

47. Yamamoto, T., Nishimura, T., Suzuki, T., and Tamon, H. (2001). Control of mesoporosity of carbon gels prepared by sol-gel polycondensation and freeze drying, *J. Non-Cryst. Solids*, **288**, pp. 46–55.

48. Shen, J., Hou, J.Q., Guo, Y.Z., Xue, H., Wu, G.M., and Zhou, B. (2005). Microstructure control of RF and carbon aerogels prepared by sol-gel process, *J.Sol-Gel Sci. Technol.*, **36**, pp. 131–136.

49. Fischer, U., Saliger, R., Bock, V., Petricevic, R., and Fricke, J. (1997). Carbon aerogels as electrode material in supercapacitors, *J. Porous Mater.*, **4**, pp. 281–285.

50. Lu, X.P., Nilsson, O., Fricke, J., and Pekala, R.W. (1993). Thermal and electrical-conductivity of monolithic carbon aerogels, *J. Appl. Phys.*, **73**, pp. 581–584.

51. Pekala, R.W., and Alviso, C.T. (1992). Carbon aerogels and xerogels, *MRS Proc.*, **270**, pp. 3–14.

52. Wen, Z.B., Qu, Q.T., Gao, Q., Zheng, X.W., Hu, Z.H., Wu, Y.P., Liu, Y.F., and Wang, X.J. (2009). An activated carbon with high capacitance

from carbonization of a resorcinol-formaldehyde resin, *Electrochem. Commun.*, **11**, pp. 715–718.

53. Conceição, F.L., Carrott, P.J.M., and Ribeiro Carrott, M.M.L. (2009). New carbon materials with high porosity in the 1–7 nm range obtained by chemical activation with phosphoric acid of resorcinol-formaldehyde aerogels, *Carbon*, **47**, pp. 1874–1877.
54. Baumann, T.F., Worsley, M.A., Han, T.Y.J., and Satcher, J.H. (2008). High surface area carbon aerogel monoliths with hierarchical porosity, *J. Non-Cryst. Solids*, **354**, pp. 3513–3515.
55. Feaver, A., and Cao, G.Z. (2006). Activated carbon cryogels for low pressure methane storage, *Carbon*, **44**, pp. 590–593.
56. Wu, D., Sun, Z., and Fu, R. (2006). Structure and adsorption properties of activated carbon aerogels, *J. Appl. Polym. Sci.*, **99**, pp. 2263–2267.
57. Lin, C., and Ritter, J.A. (2000). Carbonization and activation of sol-gel derived carbon xerogels, *Carbon*, **38**, pp. 849–861.
58. Saliger, R., Fischer, U., Herta, C., and Fricke, J. (1998). High surface area carbon aerogels for supercapacitors, *J. Non-Cryst. Solids*, **225**, pp. 81–85.
59. Hanzawa, Y., Kaneko, K., Pekala, R.W., and Dresselhaus, M.S. (1996). Activated carbon aerogels, *Langmuir*, **12**, pp. 6167–6169.
60. Steiner, S.A., Baumann, T.F., Kong, J., Satcher, J.H., and Dresselhaus, M.S. (2007). Iron-doped carbon aerogels: novel porous substrates for direct growth of carbon nanotubes, *Langmuir*, **23**, pp. 5161–5166.
61. Fu, R.W., Baumann, T.F., Cronin, S., Dresselhaus, G., Dresselhaus, M.S., and Satcher, J.H. (2005). Formation of graphitic structures in cobalt- and nickel-doped carbon aerogels, *Langmuir*, **21**, pp. 2647–2651.
62. Baker, W.S., Long, J.W., Stroud, R.M., and Rolison, D.R. (2004). Sulfur-functionalized carbon aerogels: a new approach for loading high-surface-area electrode nanoarchitectures with precious metal catalysts, *J. Non-Cryst. Solids*, **350**, pp. 80–87.
63. Marie, J., Berthon-Fabry, S., Achard, P., Chatenet, M., Pradourat, A., and Chainet, E. (2004). Highly dispersed platinum on carbon aerogels as supported catalysts for PEM fuel cell-electrodes: comparison of two different synthesis paths, *J. Non-Cryst. Solids*, **350**, pp. 88–96.
64. Baumann, T.F., Fox, G.A., Satcher, J.H., Yoshizawa, N., Fu, R.W., and Dresselhaus, M.S. (2002). Synthesis and characterization of copper-doped carbon aerogels, *Langmuir*, **18**, pp. 7073–7076.
65. Worsley, M.A., Olson, T.Y., Lee, J.R.I., Willey, T.M., Nielsen, M.H., Roberts, S.K., Pauzauskie, P.J., Biener, J., Satcher, J.H., and Baumann, T.F. (2011).

High surface area, sp^2-cross-linked three-dimensional graphene monoliths, *J. Phys. Chem. Lett.*, **2**, pp. 921–925.

66. Worsley, M.A., Pauzauskie, P.J., Olson, T.Y., Biener, J., Satcher, J.H., and Baumann, T.F. (2010). Synthesis of graphene aerogel with high electrical conductivity, *J. Am. Chem. Soc.*, **132**, pp. 14067–14069.
67. Worsley, M.A., Kucheyev, S.O., Satcher, J.H., Hamza, A.V., and Baumann, T.F. (2009). Mechanically robust and electrically conductive carbon nanotube foams, *Appl. Phys. Lett.*, **94**, p. 073115.
68. Worsley, M.A., Pauzauskie, P.J., Kucheyev, S.O., Zaug, J.M., Hamza, A.V., Satcher, J.H., and Baumann, T.F. (2009). Properties of single-walled carbon nanotube-based aerogels as a function of nanotube loading, *Acta Mater.*, **57**, pp. 5131–5136.
69. Worsley, M.A., Satcher, J.H., and Baumann, T.F. (2009). Enhanced thermal transport in carbon Aerogel nanocomposites containing double-walled carbon nanotubes, *J. Appl. Phys.*, **105**, p. 084316.
70. Worsley, M.A., Satcher, J.H., and Baumann, T.F. (2008). Synthesis and characterization of monolithic carbon aerogel nanocomposites containing double-walled carbon nanotubes, *Langmuir*, **24**, pp. 9763–9766.
71. Bordjiba, T., Mohamedi, M., and Dao, L.H. (2007). Synthesis and electrochemical capacitance of binderless nanocomposite electrodes formed by dispersion of carbon nanotubes and carbon aerogels, *J. Power Sources*, **172**, pp. 991–998.
72. Tao, Y., Noguchi, D., Yang, C.-M., Kanoh, H., Tanaka, H., Yudasaka, M., Iijima, S., and Kaneko, K. (2007). Conductive and mesoporous single-wall carbon nanohorn/organic aerogel composites, *Langmuir*, **23**, pp. 9155–9157.
73. Worsley, M.A., Satcher, J.H., and Baumann, T.F. (2010). Influence of sodium dodecylbenzene sulfonate on the structure and properties of carbon aerogels, *J. Non-Cryst. Solids*, **356**, pp. 172–174.
74. Gross, A.F., and Nowak, A.P. (2010). Hierarchical carbon foams with independently tunable mesopore and macropore size distributions, *Langmuir*, **26**, pp. 11378–11383.
75. Lu, A.H., Spliethoff, B., and Schuth, F. (2008). Aqueous synthesis of ordered mesoporous carbon via self-assembly catalyzed by amino acid, *Chem. Mater.*, **20**, pp. 5314–5319.
76. Wu, D., Fu, R., Dresselhaus, M.S., and Dresselhaus, G. (2006). Fabrication and nano-structure control of carbon aerogels via a microemulsion-templated sol-gel polymerization method, *Carbon*, **44**, pp. 675–681.

77. Baumann, T.F., and Satcher, J.H. (2004). Template-directed synthesis of periodic macroporous organic and carbon aerogels, *J. Non-Cryst. Solids*, **350**, pp. 120–125.

78. Baumann, T.F., and Satcher, J.H. (2003). Homogeneous incorporation of metal nanoparticles into ordered macroporous carbons, *Chem. Mater.*, **15**, pp. 3745–3747.

79. Lee, K.T., and Oh, S.M. (2002). Novel synthesis of porous carbons with tunable pore size by surfactant-templated sol-gel process and carbonisation, *Chem. Commun.*, pp. 2722–2723.

80. Marie, J., Berthon-Fabry, S., Chatenet, M., Chainet, E., Pirard, R., Cornet, N., and Achard, P. (2007). Platinum supported on resorcinol-formaldehyde based carbon aerogels for pemfc electrodes: influence of the carbon support on electrocatalytic properties, *J. Appl. Electrochem.*, **37**, pp. 147–153.

81. Sassin, M.B., Mansour, A.N., Pettigrew, K.A., Rolison, D.R., and Long, J.W. (2010). Electroless deposition of conformal nanoscale iron oxide on carbon nanoarchitectures for electrochemical charge storage, *ACS Nano*, **4**, pp. 4505–4514.

82. Long, J.W., Sassin, M.B., Fischer, A.E., Rolison, D.R., Mansour, A.N., Johnson, V.S., Stallworth, P.E., and Greenbaum, S.G. (2009). Multifunctional MnO_2-carbon nanoarchitectures exhibit battery and capacitor characteristics in alkaline electrolytes, *J. Phys. Chem. C*, **113**, pp. 17595–17598.

83. Fischer, A.E., Saunders, M.P., Pettigrew, K.A., Rolison, D.R., and Long, J.W. (2008). Electroless Deposition of nanoscale MnO_2 on ultraporous carbon nanoarchitectures: correlation of evolving pore-solid structure and electrochemical performance, *J. Electrochem. Soc.*, **155**, pp. A246–A252.

84. Fischer, A.E., Pettigrew, K.A., Rolison, D.R., Stroud, R.M., and Long, J.W. (2007). Incorporation of homogeneous, nanoscale MnO_2 within ultraporous carbon structures via self-limiting electroless deposition: implications for electrochemical capacitors, *Nano Lett.*, **7**, pp. 281–286.

85. Worsley, M.A., Kuntz, J.D., Satcher, J.H., and Baumann, T.F. (2010). Synthesis and characterization of monolithic, high surface area SiO_2/C and SiC/C composites, *J. Mater. Chem.*, **20**, pp. 4840–4844.

86. Worsley, M.A., Kuntz, J.D., Pauzauskie, P.J., Cervantes, O., Zaug, J.M., Gash, A.E., Satcher, J.H., and Baumann, T.F. (2009). High surface area carbon nanotube-supported titanium carbonitride aerogels, *J. Mater. Chem.*, **19**, pp. 5503–5506.

87. Worsley, M.A., Kuntz, J.D., Cervantes, O., Han, T.Y.J., Gash, A.E., Satcher, J.H., and Baumann, T.F. (2009). Route to high surface area TiO_2/C and TiCN/C composites, *J. Mater. Chem.*, **19**, pp. 7146–7150.

88. Litster, S., and McLean, G. (2004). PEM fuel cell electrodes, *J. Power Sources*, **130**, pp. 61–76.

89. Glora, M., Wiener, M., Petricevic, R., Probstle, H., and Fricke, J. (2001). Integration of carbon aerogels in PEM fuel cells, *J. Non-Cryst. Solids*, **285**, pp. 283–287.

90. Petricevic, R., Glora, M., and Fricke, J. (2001). Planar fibre reinforced carbon aerogels for application in PEM fuel cells, *Carbon*, **39**, pp. 857–867.

91. Queheillalt, D.T., Hass, D.D., Sypeck, D.J., and Wadley, H.N.G. (2001). Synthesis of open-cell metal foams by templated directed vapor deposition, *J. Mater. Res.*, **16**, pp. 1028–1036.

92. George, S.M. (2010). Atomic layer deposition: an overview, *Chem. Rev.*, **110**, pp. 111–131.

93. Baumann, T.F., Biener, J., Wang, Y.M.M., Kucheyev, S.O., Nelson, E.J., Satcher, J.H., Elam, J.W., Pellin, M.J., and Hamza, A.V. (2006). Atomic layer deposition of uniform metal coatings on highly porous aerogel substrates, *Chem. Mater.*, **18**, pp. 6106–6108.

94. Biener, J., Baumann, T.F., Wang, Y.M., Nelson, E.J., Kucheyev, S.O., Hamza, A.V., Kemell, M., Ritala, M., and Leskela, M. (2007). Ruthenium/aerogel nanocomposites via atomic layer deposition, *Nanotechnology*, **18**, p. 055303.

95. Kucheyev, S.O., Biener, J., Baumann, T.F., Wang, Y.M., Hamza, A.V., Li, Z., Lee, D.K., and Gordon, R.G. (2008). Mechanisms of atomic layer deposition on substrates with ultrahigh aspect ratios, *Langmuir*, **24**, pp. 943–948.

96. Ghosal, S., Baumann, T.F., King, J.S., Kucheyev, S.O., Wang, Y.M., Worsley, M.A., Biener, J., Bent, S.F., and Hamza, A.V. (2009). Controlling atomic layer deposition of TiO_2 in aerogels through surface functionalization, *Chem. Mater.*, **21**, pp. 1989–1992.

97. Kucheyev, S.O., Biener, J., Wang, Y.M., Baumann, T.F., Wu, K.J., van Buuren, T., Hamza, A.V., Satcher, J.H., Elam, J.W., and Pellin, M.J. (2005). Atomic layer deposition of ZnO on ultralow-density nanoporous silica aerogel monoliths, *Appl. Phys. Lett.*, **86**, p. 083108.

98. Aaltonen, T., Ritala, M., Sajavaara, T., Keinonen, J., and Leskela, M. (2003). Atomic layer deposition of platinum thin films, *Chem. Mater.*, **15**, pp. 1924–1928.

99. Aaltonen, T., Alen, P., Ritala, M., and Leskela, M. (2003). Ruthenium thin films grown by atomic layer deposition, *Chem. Vap. Depos.*, **9**, pp. 45–49.

100. Liu, C., Wang, C.C., Kei, C.C., Hsueh, Y.C., and Perng, T.P. (2009). Atomic layer deposition of platinum nanoparticles on carbon nanotubes for application in proton-exchange membrane fuel cells, *Small*, **5**, pp. 1535–1538.

101. Worsley, M.A., Stadermann, M., Wang, Y.M., Satcher, J.H., and Baumann, T.F. (2010). High Surface area carbon aerogels as porous substrates for direct growth of carbon nanotubes, *Chem. Commun.*, **46**, pp. 9253–9255.

Chapter 4

Carbon Electronics

Colin Johnston
Department of Materials, University of Oxford, Begbroke Science Park, Sandy Lane, Yarnton, Oxford, OX5 1PF, U.K.
colin.johnston@materials.ox.ac.uk

Carbon has been investigated as an electronic material for many years. By far the bulk of research has concentrated on diamond, although other forms of carbon, such as diamond-like carbon (DLC), carbon nanotubes (CNTs), graphene, and fullerenes, have attracted considerable attention in more recent years. Yet, the commercial exploitation of carbon in active electronics remains elusive.

4.1 Diamond

Early work on the electronic properties of diamond used naturally occurring diamond stones. However, the natural variability of this material limited the active electronic applications of diamond. Nevertheless, enough significant research was undertaken on the small quantity of naturally occurring electrical quality stones to highlight the potential of diamond as an active electronic material.

Carbon-based Nanomaterials and Hybrids: Synthesis, Properties, and Commercial Applications
Edited by Hans-Jörg Fecht, Kai Brühne, and Peter Gluche

ISBN 978-981-4316-85-9 (Hardcover), 978-981-4411-41-7 (eBook)
www.panstanford.com

Diamond is a wide-bandgap material (ca. 5.47 eV) with impressive properties that include high carrier mobilities, high breakdown strength, and high thermal conductivity. Table 4.1 lists some of the basic material properties of diamond and compares it to silicon.

Table 4.1 Comparison of relevant electrical properties

Property	Diamond	Silicon
Bandgap (eV)	5.47	1.1
Electron carrier mobility ($cm^2V^{-1}s^{-1}$)	2,200	1,500
Saturated electron velocity ($\times 10^6$ cm s^{-1})	27	10
Hole carrier mobility ($cm^2V^{-1}s^{-1}$)	1,600	600
Breakdown strength (MV cm^{-1})	10	0.3
Dielectric constant	5.5	11.8
Resistivity (Ω cm)	1013	103
Thermal conductivity (W $cm^{-1}K^{-1}$)	20	1.5
Johnson's figure of merit (normalized to silicon)	8200 [1]	1
Keyes's figure of merit (normalized to silicon)	32 [1]	1

These extreme properties of diamond make it very attractive for high-power and high-frequency electronics and applications in particular, whilst the wide bandgap makes it an attractive material for high-temperature applications. Two figures of merit can be used to compare materials for these applications:

(i) Johnson's figure of merit is a measure of suitability of a semiconductor material for high-frequency power transistor applications and requirements and is applicable to both field-effect transistors (FETs) and, with proper interpretation of the parameters, also to bipolar-junction transistors. It is the product of the charge carrier saturation velocity in the material and the electric breakdown field under same conditions, first proposed by A. Johnson in 1965 [2].

(ii) Keyes's figure of merit (Keyes, 1972 [3]) takes into account the thermal properties of a material and assumes that smaller devices are inherently faster in response to a fixed input but smaller devices have higher thermal resistances that limit their power output, hence introducing thermal conductivity as an optimization parameter.

Yet despite these attractive figures of merit diamond device technology development has been somewhat limited. Initially, this was due to a lack of availability of suitable natural stones, but even with the ready availability of chemical vapor deposition (CVD) diamond films, commercial success of diamond electronic devices remains elusive.

Dopants in wide-bandgap semiconductors tend to have higher ionization energies than in narrow-bandgap semiconductors, resulting in low activation at room temperature. This lack of shallow dopants in diamond has led some investigators [4] to conclude that the prospects of using diamond for electronic devices are poor and that conventional device designs cannot automatically be expected to work well in diamond. Specific approaches need to be adopted that exploit the material advantages offered by diamond rather than to adopt conventional device designs. Examples of this approach are delta-doped FETs [5] and Schottky diodes with intrinsic layers [6]. In these types of devices, the charge carriers are supplied from a highly doped region, while the active region consists of undoped diamond where carrier mobilities are high.

For devices operating at elevated temperatures high dopant ionization energy is less of a problem and conventional designs work better. Yet, this application area suffers from competition from modified silicon and the other wide-bandgap semiconductors silicon carbide and gallium nitride.

Boron (B), which is an acceptor in diamond, has an ionization energy of 0.37 eV, and extrinsic p-type conduction is observed, although at high dopant concentrations metallic behavior dominates. Figure 4.1 shows the activation energy as a function of effective dopant concentration for boron in single-crystal polycrystalline CVD diamond.

The actual current conduction mechanism depends on the doping level. In the case of B-doped diamond two important conduction mechanisms working in parallel must be considered, that is, impurity band and band conduction. At low doping concentrations, band conduction dominates, and at high doping levels, impurity band conduction dominates. The impurity band increases with increasing doping concentration. When the impurity band merges with the valence band, edge metallic behavior is observed.

Whilst p-type dopants in diamond are relatively well understood, n-type dopants are still problematic. The group V [8] elements have been shown to impart n-type behavior when doped into diamond. Nitrogen is the most extensively studied of these dopants, but its activation energy is very high at ca. 1.7 eV. Moving down the periodic table, phosphorus, when introduced into the diamond lattice, has an activation energy of about 0.6 eV, and arsenic exhibits activation energies of about 0.3 eV, although these much large atoms are harder to introduce into the lattice in a controlled and reproducible manner. The group I element lithium has also shown to be an n-type dopant, and work at the Harwell Laboratories in the 1970–1980s [9] showed that ion implantation could be used to control doping culminating in a p-n transistor device fabricated by Li ion implantation in a naturally B-doped (type 2B) diamond stone.

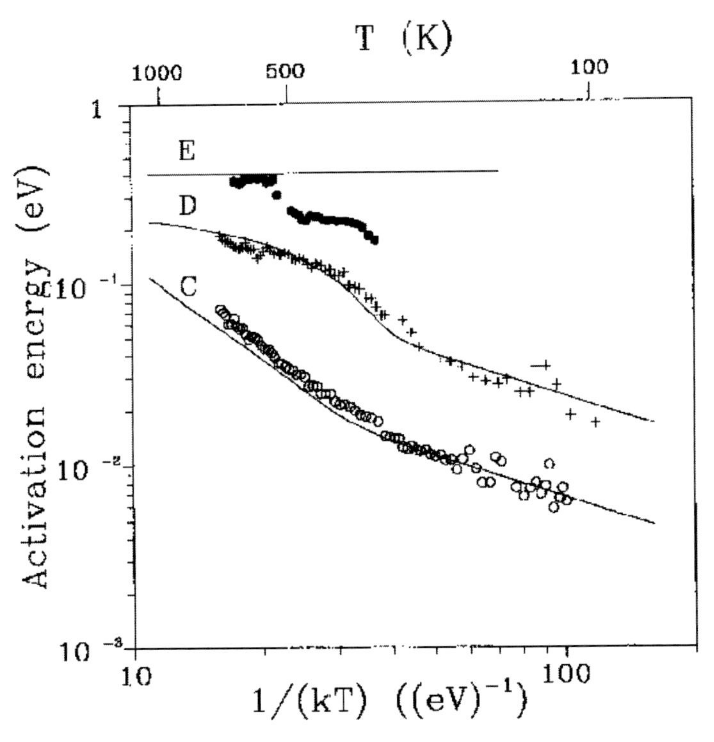

Figure 4.1 Activation energy for single-crystalline (filled symbols) and polycrystalline diamond (open symbols) vs. effective doping concentration (Reprinted from Ref. [7] by permission of John Wiley and Sons. Copyright Â© 1996 WILEY-VCH Verlag GmbH & Co. KGaA).

Alkali metals of a smaller atomic size may be good n-type dopants since it is known from nuclear reaction studies [10] that lithium occupies an interstitial site in the diamond lattice and can therefore donate an electron to the conduction band.

Figure 4.2 shows the variation in sheet resistance of the lithium-implanted layer as a function of lithium dose. There is a strong dependence of sheet resistance on the lithium dose, and implantation produces a lower resistance at 800°C than at 120°C, which is probably due to the in situ annealing out, at the higher temperature, of defects in the diamond, which would otherwise contribute to conduction.

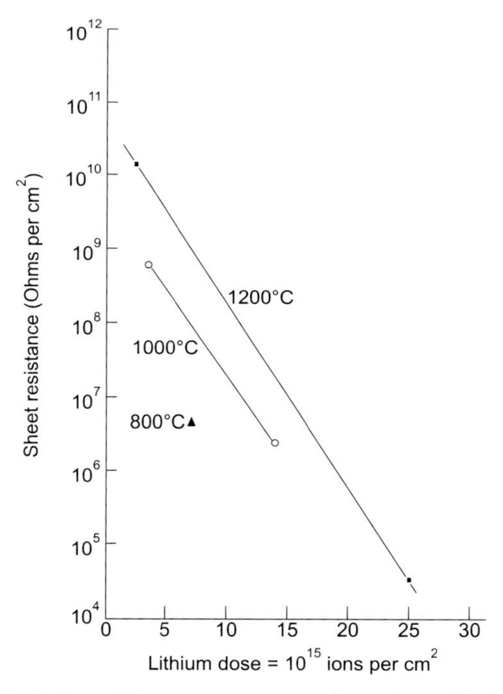

Figure 4.2 Variation of sheet resistance with implanted lithium dose for three temperatures of implantation [Reprinted from Ref. [9], Copyright (1991), with permission from Elsevier].

Figure 4.3 shows the *I–V* characteristic of the junction produced at Harwell, which was good at room temperature and retained excellent properties at temperatures up to 350°C with excellent reverse bias breakdown characteristics and low leakage currents.

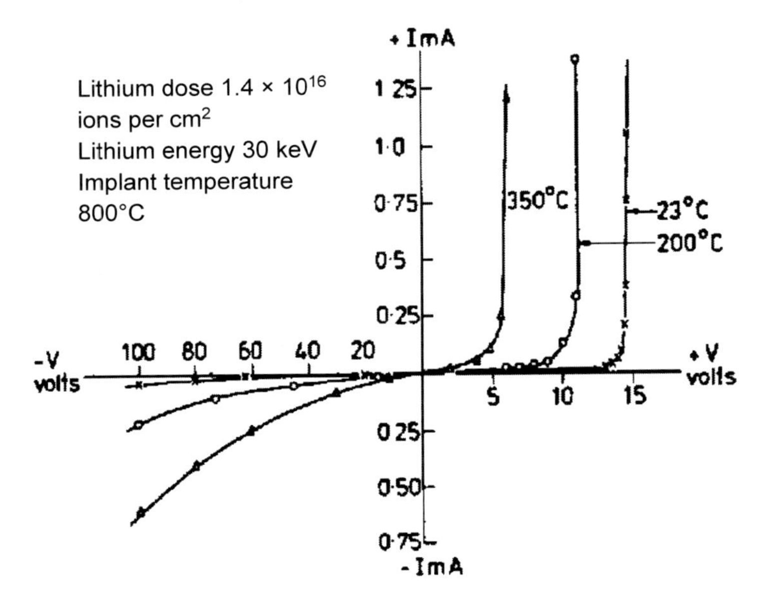

Figure 4.3 The p-n junction in type 2B diamond produced by lithium ion implantation [Reprinted from Ref. [9], Copyright (1991), with permission from Elsevier.]

In fact, the Harwell p-n junction was not the first to be demonstrated. Prins [11] fabricated a bipolar n-p-n transistor by carbon ion implantation of a type 2 diamond, thus creating zones of high defect concentrations that exhibited n-type behavior.

Unipolar devices have been demonstrated the most extensively, and a short peruse of the most recent Diamond and Related Materials Conferences [12] and others [13] shows that most diamond device research is based on FETs: metal–insulator–semiconductor FET (MISFET) and metal–semiconductor FET (MESFET) and Schottky diodes.

In a conventional FET device, current transport takes place in a channel below the gate between the source and drain contacts. The channel current is limited to

$$I_{\text{channel}} = q \cdot \rho \cdot d \cdot v_{\text{sat}}, \tag{4.1}$$

where ρ is the hole concentration, d is the channel thickness, and v_{sat} the hole saturation velocity. To reach a high current, a high carrier concentration is necessary, which, due to the high ionization energy of boron-doped diamond, means that very high doping concentrations are required. This results in material with a high concentration of scattering centers and, consequently, poor transport properties. Thus, other device designs that make use of the superior transport properties of intrinsic diamond have been explored, including:

(i) "surface transfer doping," where a hole channel is created near the surface by hydrogen termination of the diamond surface to produce superior performance, although the long-term stability of such surface-transfer-doped FETs remains an issue, especially when these devices are used at high temperatures [14–16].

(ii) "delta doping" to achieve a spatial separation between ionized acceptors and holes. The delta layer is a thin (a few nanometers), highly doped layer surrounded by intrinsic diamond. Conduction mainly occurs in the intrinsic layers (with good transport properties), while the charge carriers are supplied by diffusion from the narrow delta-doped layer (where ionization is almost complete). Some progress has been made in producing prototype delta-doped devices with real commercial potential [17, 18].

To achieve the highest performance in power diodes, one requires a high breakdown voltage (V_{br}) together with low resistance, but these depend on the doping concentration. Diamond is attractive since it has high breakdown strength and high hole mobility, but since boron is incompletely activated due to its relatively high activation energy, device performance is compromised. Therefore, other device structures have been investigated, including metal-insulator-p+ diodes where an intrinsic layer lies between the Schottky metal contact and the B-doped layer. Furthermore, if the doping concentration is high enough in this layer, conduction is metallic in nature and complete activation is achieved and the intrinsic layer is space charge limiting, and good performance has been demonstrated [19, 20].

Specialized devices have also been extensively demonstrated in diamond, including various sensors, microelectrodes, ultraviolet (UV) detectors and photodiodes, ionization detectors, and microelectromechanical systems (MEMS). Perhaps the most overlooked application of diamond in electronics is as a thermal management substrate where both natural diamond and CVD diamond are unsurpassed for their performance benefits. Diamond thermal management substrates currently provide the largest commercial market in the electronics sector.

4.2 Diamond-Like Carbon

A few active electronic applications of diamond-like carbon (DLC) have been demonstrated, including electroluminescent devices and MISFET structures, although performance has been generally very limited due to generally low resistivity of the DLC and the large defect concentrations present in these films or their interfaces that act as carrier traps [21]. Again, however, few of the demonstrated devices have been commercialized. DLC has found commercial application in electronics but only in a passive role, where, for example, it can be used to passivate edge terminations in large silicon power diodes.

More success [22] has been achieved with so-called tetrahedral amorphous carbon (ta-C). In ta-C, it is the π bonding (can be thought of as the valence band) and π^* antibonding (conduction band) electronic states resulting from p-orbital bonding at sp^2 sites that determines the optical bandgap. In ta-C with a very low sp^3 content, the π and π^* electronic states are highly localized within the much wider-gap (ca. 5 eV) σ–σ^* states arising mainly from the sp^3-hybridized diamond-like bonds. The best electronic properties for ta-C have been obtained from material comprising 30–40% sp^2 bonds. This material has a bandgap of ca. 2 eV, and electronic doping with P and N has been achieved, allowing the conductivity to be tailored, as shown in Fig. 4.4.

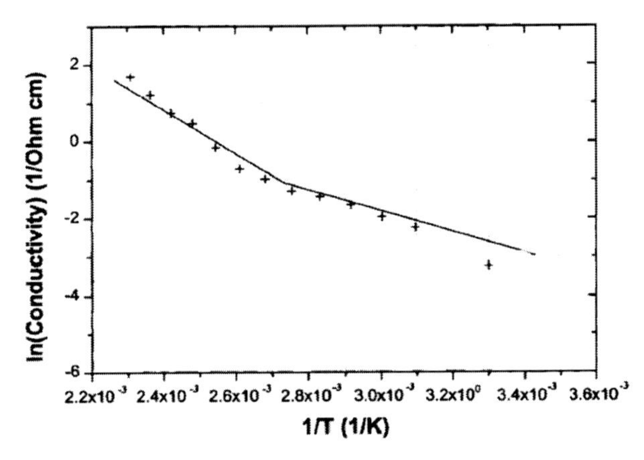

Figure 4.4 Conductivity dependence with temperature for ta-C:N [Reprinted from Ref. [22], Copyright (1998), with permission from Elsevier].

4.3 Carbon Nanotubes

Carbon nanotubes (CNTs) are in effect tubes of carbon atoms in a predominantly sp^2 bonding configuration of some nanometers in diameter and many microns long, comprising either a single-walled CNT (SWCNT) or a multiple-walled CNT (MWCNT) with graphene-like conduction along the axis of the tube, which can be metallic or semiconducting. Devices fabricated from CNTs are in effect molecular devices with a single tube acting as a single molecule. Therefore, to fabricate effective devices one must manipulate individual tubes or small entangled clusters of tubes, which is nontrivial. Nevertheless, since their discovery [23] in the early 1990s there has been phenomenal worldwide activity investigating their electrical and electronic properties and their potential applications in electronics.

McEuen et al. [24] elegantly demonstrate the semiconducting nature of an SWCNT by constructing a metal-oxide–semiconductor FET (MOSFET) analog by contacting a source and a drain contact through a single SWCNT on a conducting substrate, which acts as the gate electrode. Figure 4.5 [24] shows the measurement of the conductance as the gate voltage (V_g) is varied. The tube conducts at negative V_g and turns off with a positive V_g. The resistance change between the on and off states is many orders of magnitude.

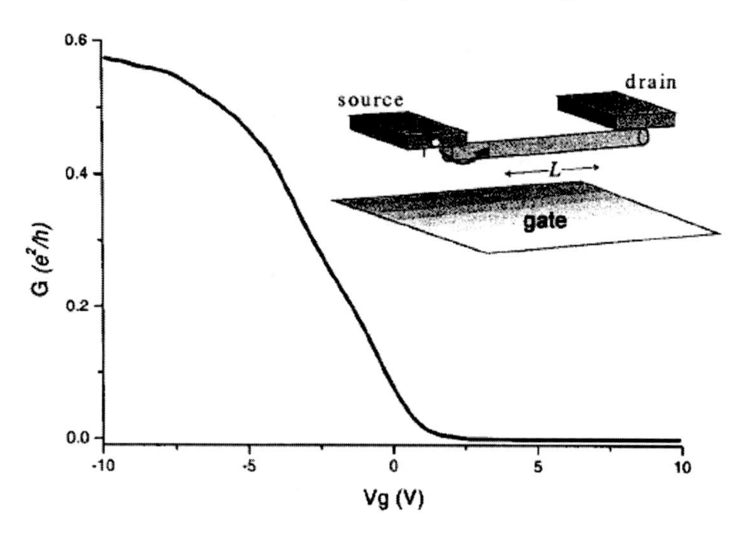

Figure 4.5 Conductance (G) versus gate voltage (V_g) for a p-type SWCNT FET. The inset shows the device geometry schematically.

Semiconducting nanotubes are typically p-type, most likely due to chemical species, particularly oxygen, adsorbing on the tube that act as weak p-type dopants. Adsorbate doping can be a problem for reproducible device behavior; however, controlled substitutional doping of tubes, both p- and n-type, has been accomplished in a number of ways, primarily by using boron for p-type doping and nitrogen for n-type doping, substituting a carbon atom in the graphitic wall structure. The semiconducting behavior is demonstrated by plotting the temperature dependence of resistance of the doped CNTs, as shown in Fig. 4.6 [25].

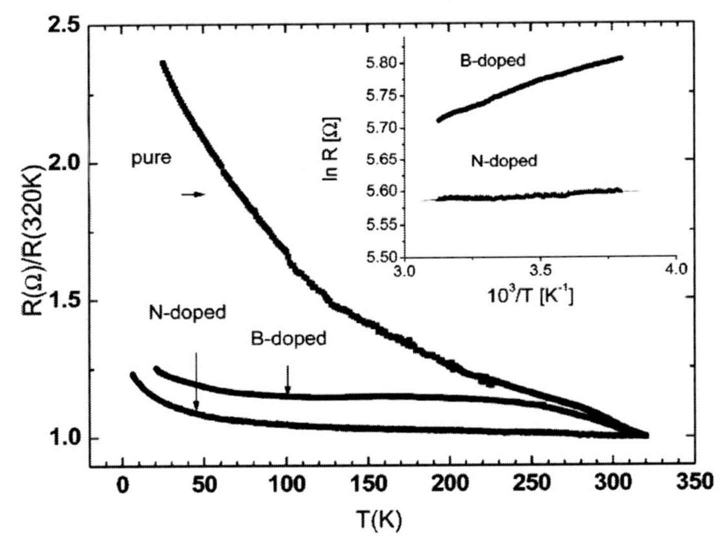

Figure 4.6 Two-probe measurements of the temperature dependence of R for pristine, B-doped, and N-doped nanotube mats (Reprinted with permission from Ref. [25]. Copyright (2003) American Chemical Society).

Numerous designs for FET structures can be found in the literature, including those shown schematically in Fig. 4.7.

The earliest techniques for fabricating CNT FETs involved prepatterning parallel strips of metal across a silicon dioxide substrate and then depositing the CNTs on top in a random pattern [26]. The semiconducting CNTs that happened to fall across two metal strips meet all the requirements necessary for a rudimentary FET. One metal strip is the "source" contact, while the other is the "drain" contact. The silicon dioxide substrate can be used as the gate oxide, and adding a metal contact on the back makes the semiconducting CNT gateable.

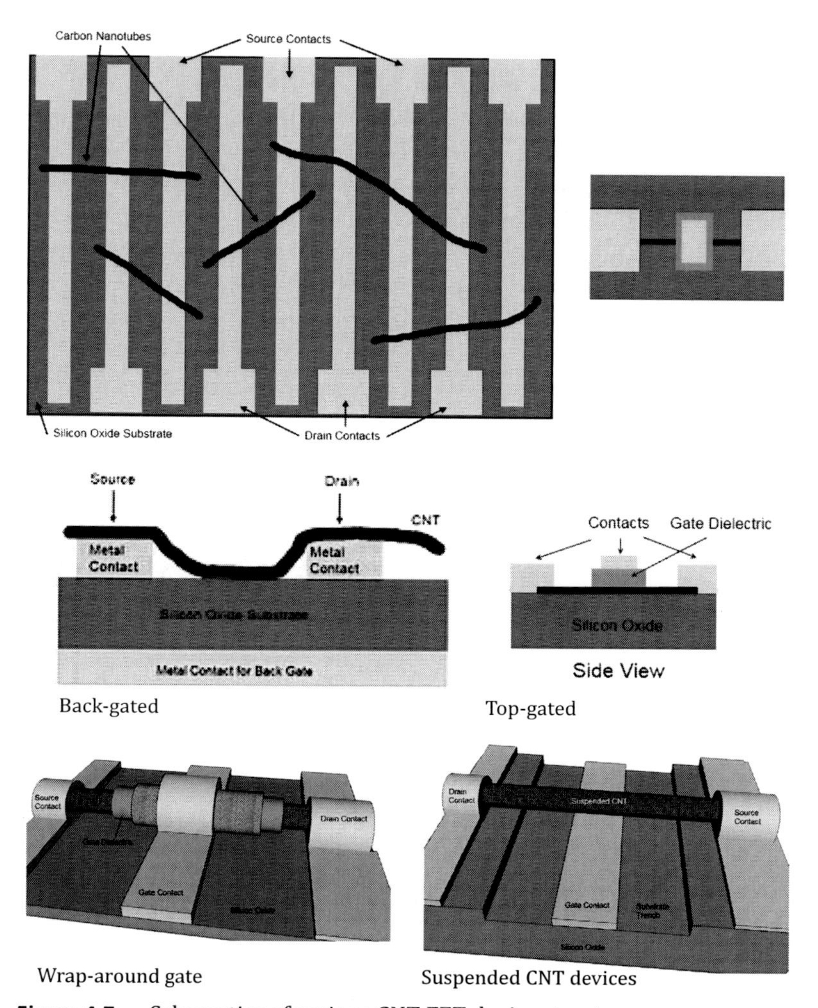

Figure 4.7 Schematics of various CNT FET device structures.

This technique suffers from several drawbacks, which make for nonoptimized transistors, the first being the metal contact, which actually has very little contact with the CNT, since the CNT just lies on top of it and the contact area is therefore very small. Also, due to the semiconducting nature of the CNT, a Schottky barrier forms at the metal–semiconductor interface, [27] increasing the contact resistance. The second drawback is due to the back-gate device geometry. Its thickness makes it difficult to switch the devices on and off using low voltages, and the fabrication process led to poor contact between the gate dielectric and the CNT [28].

Eventually, researchers migrated from the back-gate approach to a more advanced top-gate fabrication process [28]. In the first step, CNTs are solution-deposited onto a silicon oxide substrate. Individual CNTs are then located via atomic force microscopy (AFM) or scanning electron microscopy. After an individual tube is isolated, source and drain contacts are defined and patterned using high-resolution electron beam lithography. A high-temperature anneal step reduces the contact resistance by improving adhesion between the contacts and the CNT. A thin top-gate dielectric is then deposited on top of the CNT, either via evaporation or via atomic layer deposition. Finally, the top-gate contact is deposited on the gate dielectric, completing the process.

Arrays of top-gated CNT FETs can be fabricated on the same wafer, since the gate contacts are electrically isolated from each other, unlike in the back-gated case. Also, due to the thinness of the gate dielectric, a larger electric field can be generated with respect to the CNT using a lower gate voltage. These advantages mean top-gated devices are generally preferred over back-gated CNT FETs, despite their more complex fabrication process.

Wrap-around-gate CNT FETs, also known as gate-all-around CNT FETs, were developed in 2008 [29] and are a further improvement upon the top-gate device geometry. In this device, instead of gating just the part of the CNT that is closer to the metal gate contact, the entire circumference of the CNT is gated. This should ideally improve the electrical performance of the CNT FET, reducing leakage current and improving the device on/off ratio.

Device fabrication begins by first wrapping CNTs in a gate dielectric and gate contact via atomic layer deposition [30]. These wrapped CNTs are then solution-deposited on an insulating substrate, where the wrappings are partially etched off, exposing the ends of the CNTs. The source, drain, and gate contacts are then deposited onto the CNT ends and the metallic outer gate wrapping.

Yet another CNT FET device geometry involves suspending the CNT over a trench to reduce contact with the substrate and gate oxide [31]. This technique has the advantage of reduced scattering at the CNT–substrate interface, improving device performance [29, 32, 33]. There are many methods used to fabricate suspended CNT FETs, ranging from growing them over trenches using catalyst particles [31], transferring them onto a substrate, and then underetching the dielectric beneath [32] and transfer-printing onto a trenched substrate [33].

The main problem suffered by suspended CNT FETs is that they have very limited material options for use as a gate dielectric (generally air or vacuum), and applying a gate bias has the effect of pulling the CNT closer to the gate, which places an upper limit on how much the CNT can be gated. This technique will also only work for shorter CNTs, as longer tubes will flex in the middle and droop toward the gate, possibly touching the metal contact and shorting the device. In general, suspended CNT FETs are not practical for commercial applications, but they can be useful for studying the intrinsic properties of clean CNTs.

One can understand the interest in CNT FET structures, given the startingly high conductances and mobilities that have been achieved in single CNTs, as shown, for example, by Fuhrer's group at the University of Maryland. Figure 4.8 is reproduced from their website [34], which shows the conductance of a long semiconducting CNT strung between two gold electrodes and mobility as a function of gate voltage. At a low gate voltage (low charge carrier density) the mobility exceeds that of InSb (77,000 cm2/Vs), the previous highest-known mobility at room temperature.

p-n junction devices have also been realized with CNTs. Mueller et al. [35] have demonstrated efficient narrow-band light emission from a single CNT p-n diode. The p- and n-type regions in a CNT diode are formed using electrostatic doping. Two separate gate electrodes, which couple to two different regions of an SWCNT, are used as can be seen in the schematic in Fig. 4.9. One gate is biased with a negative voltage, drawing holes into the nanotube channel, and the other gate is biased with a positive voltage, resulting in an accumulation of electrons in the channel, thus forming a p-n junction. The devices behave very much like conventional semiconductor diodes.

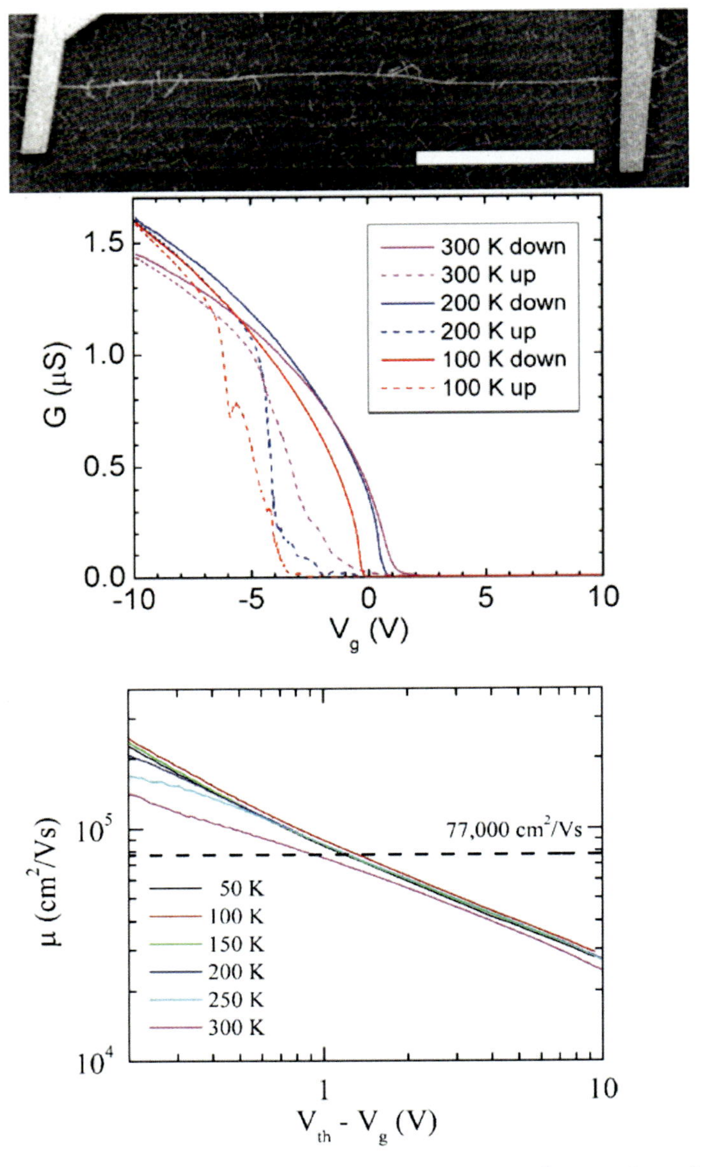

Figure 4.8 Conductance and mobility as a function of gate voltage for a single long semiconducting CNT strung between two gold electrodes shown in the top micrograph (note the scale bar is 100 μm). The substrate is silicon and acts as the back gate.

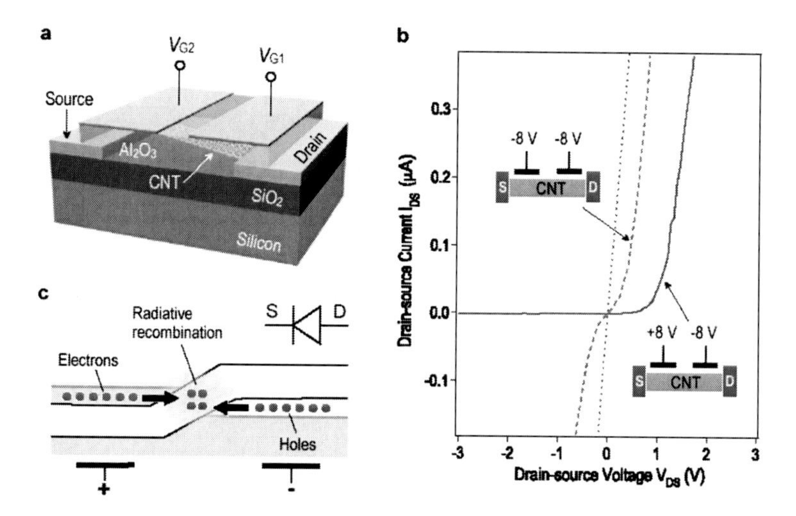

Figure 4.9 (a) Schematic of a CNT p–n diode and (b) diode characteristics under different biasing conditions with the silicon gate grounded. (c) Band structure of a CNT diode when it is biased in the forward direction: electrons and holes are injected into the intrinsic region and recombine partially radiatively and partially nonradiatively [Reprinted by permission from Macmillan Publishers Ltd: *Nature Nanotechnology*, Ref. [35], copyright (2009)].

4.4 Graphene

Graphene [36] is a zero-gap semiconductor; for most directions in the graphene sheet, there is a bandgap, and electrons are not free to flow along those directions unless they are given extra energy. However, in certain special directions, graphene is metallic and electrons flow easily along those directions.

This ambipolar electric field effect enables the charge carriers in graphene to be tuned continuously between electrons and holes in concentrations n as high as 10^{13} cm^{-2} and their mobilities μ can exceed 15,000 $cm^2V^{-1}s^{-1}$, even under ambient conditions. Moreover, the observed mobilities only weakly depend on temperature T, with the potential to exceed the highest-measured mobilities in semiconductors. In graphene, μ remains high even at high dopant levels ($>10^{12}$ cm^{-2}) in both electrically and chemically doped devices, which translates into ballistic transport on the submicrometre scale (currently up to ~0.3 μm at 300 K).

An equally important reason for the interest in graphene is a particular unique nature of its charge carriers. Although there is nothing particularly relativistic about electrons moving around carbon atoms, their interaction with the periodic potential of graphene's honeycomb lattice gives rise to new quasiparticles that, at low energies E, are accurately described by the (2+1)-dimensional Dirac equation with an effective speed of light $vF \approx 10^6$ $m^{-1}s^{-1}$. These quasiparticles, called "massless Dirac fermions," can be seen as electrons that have lost their rest mass m_0 or as neutrinos that acquired the electron charge, opening up the possibility of new devices based on the quantum electrodynamics (QED) phenomenon. For a more complete description of the electronic and other properties of graphene, refer to Geim and Novoselov's progress review in *Nature Materials* [36].

Practical devices (FETs) have been fabricated using graphene nanoribbons (GNRs), which are essentially single layers of graphene that are cut in a particular pattern to give it certain electrical properties, and depending on how the unbonded edges are configured, they can be in either a zigzag or an armchair configuration. Calculations based on tight binding predict that zigzag GNRs are always metallic, while armchairs can be either metallic or semiconducting depending on their width. High on-off ratios have been achieved from FETs fabricated with GNRs (w ca. sub-10 nm to ~55 nm). The devices had palladium (Pd) as source/drain (S/D) metal contacts (channel length L ca. 200 nm), a p++-Si back gate, and 300 nm SiO_2 as gate dielectrics. The room-temperature on-off current switching (I_{on}/I_{off}) induced by the gate voltage increased exponentially as the GNR width decreased, with $I_{on}/I_{off} \approx 1, \approx 5, \approx 100$, and $> 10^5$ for w ca. 50 nm, ca. 20 nm, ca. 10 nm, and ca. sub-10-nm, as shown in Figs. 4.10 and 4.11 [37].

Single-electron transistors have also been fabricated from graphene, and device structures defined by dimensions of 30 nm or less reveal a confinement gap of up to 0.5 eV, demonstrating the possibility of molecular-scale electronics based on graphene. Ponomarenko et al. [38] found three basic operational regimes depending on their device dimensions. Large devices exhibit (nearly) periodic Coulomb blockade (CB) resonances that at low

T are separated by regions of zero conductance, G (Fig. 4.12). As T increases, the peaks become broader and overlap, gradually transforming into CB oscillations. The oscillations become weaker as G increases with carrier concentration or T and completely disappear for G larger than ~0.5 e^2/h because the barriers become too transparent to allow CB.

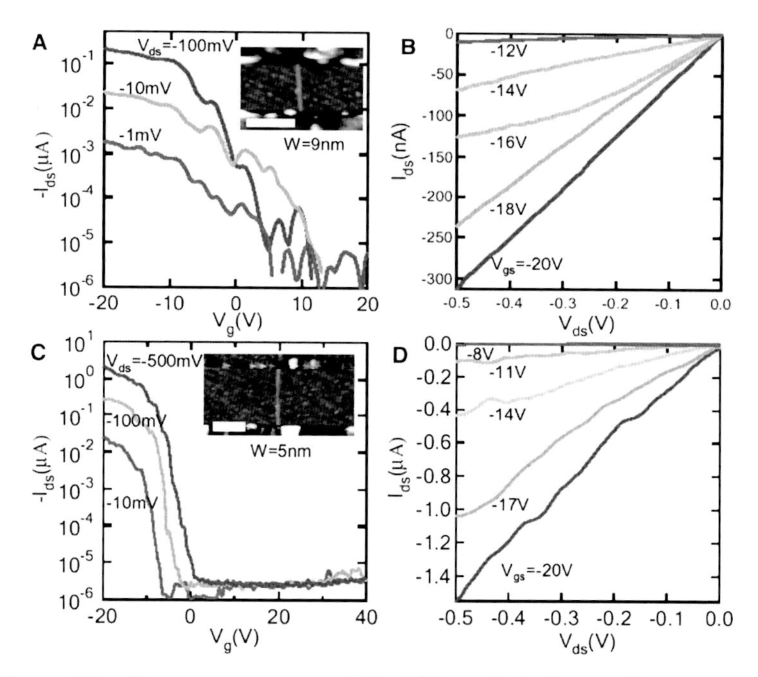

Figure 4.10 Room-temperature GNR FETs with high on-off ratios. (A) Transfer characteristics (current versus gate voltage I_{ds}–V_{gs}) for a $w \approx 9$ nm (thickness ~1.5 nm, ~2 layers) and channel length $L \approx 130$ nm GNR with Pd contacts and a Si back gate. The inset shows the AFM image of this device. Scale bar is 100 nm. (B) Current–voltage (I_{ds}–V_{ds}) curves recorded under various V_{gs} for the device in (A). (C) Transfer characteristics for a $w \approx$ 5 nm (thickness ~1.5 nm, ~2 layers) and channel length $L \approx$ 210 nm GNR with Pd contacts. The inset shows the AFM image of this device. Scale bar is 100 nm. (D) I_{ds}–V_{ds} characteristics recorded under various V_{gs} for the device in (C) (From Ref. [37]. Reprinted with permission from AAAS).

For devices smaller than ~100 nm, Ponomarenko et al. observed a qualitative change in behavior: CB peaks were no longer a periodic function of V_g but varied strongly in their spacing (as shown in Fig. 4.13), which illustrates this behavior for $D \approx 40$ nm. The size quantization becomes an important factor even for such a modest confinement.

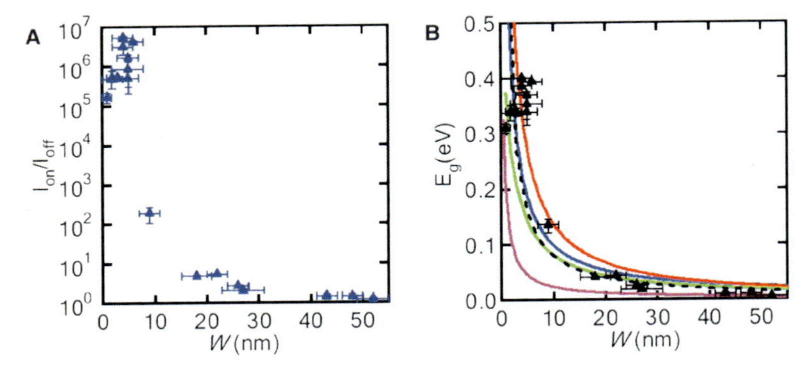

Figure 4.11 Electrical properties of GNR FETs. (A) I_{on}/I_{off} ratios (under V_{ds} = 0.5 V) for GNRs of various ribbon widths measured in this work. Error bars for the x-axis have their bases in uncertainties in the ribbon width on the basis of AFM measurements. Error bars for the y-axis have their bases in the fluctuations in the off-state current (as shown in Fig. 4.10 above). (B) E_g extracted from experimental data (symbols) for various GNRs versus ribbon width. The black dashed line is a fit to the experimental data into an empirical form of E_g (eV) = 0.8/w(nm). The purple, blue, and orange solid lines are first-principle calculations for three types of armchair-edged GNRs, respectively, and the green solid line is calculations for zigzag-edged GNRs (From Ref. [37]. Reprinted with permission from AAAS).

Yet, perhaps the property of graphene that has attracted the most recent commercial interest is its ability to function as a transparent conducting electrode required for such applications as touch screens, liquid crystal displays (LCDs), photovoltaic cells, and light-emitting diodes (LEDs), which is a passive electronic application but demands large areas of perfect graphene to be produced. It is possible that this will help to drive the commercialization of active electronic devices based on graphene more rapidly than with the other forms of carbon electronics, in a similar way that silicon carbide electronics have benefited by demand as a substrate for blue LED manufacture. This has led to a year-on-year improvement in SiC substrate quality and cost reduction as size and quantities manufactured increase.

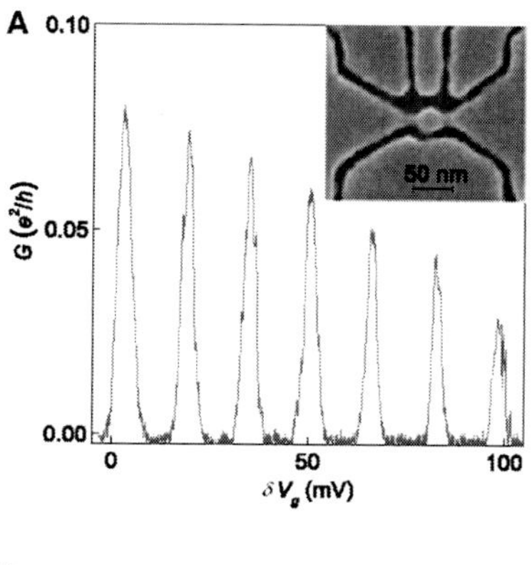

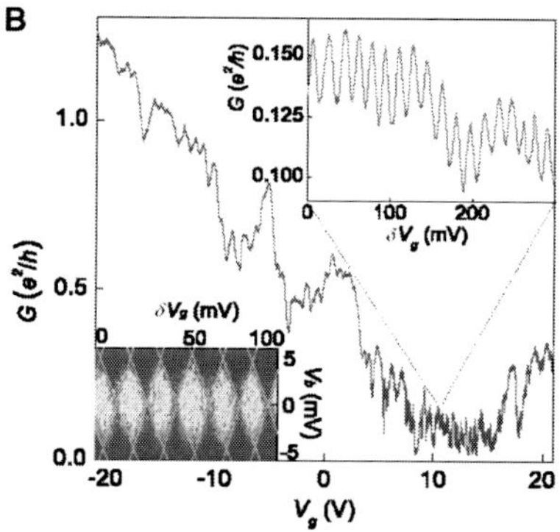

Figure 4.12 Graphene-based single-electron transistor. (A) Conductance G of a device with the central island 250 nm in diameter and distant side gates (13) as a function of V_g in the vicinity of +15 V (B); T = 0.3 K. The inset shows one of the smaller devices to illustrate the high resolution of lithography necessary to allow features down to 10 nm. Dark areas in the scanning electron micrograph are gaps in the PMMA mask so that graphene is removed from these areas by plasma etching. In this case, a 30 nm QD is connected to contact regions through narrow constrictions, and there are four side gates. (B) Conductance of the same device as in (A) over a wide range of V_g (T = 4 K). In the upper inset, zooming into the low-G region reveals hundreds of CB oscillations. The lower inset shows Coulomb diamonds: differential conductance $G_{diff} = dI/dV$ as a function of V_g (around +10 V) and bias V_b (yellow-to-red scale corresponds to G_{diff} varying from 0 to 0.3 e^2/h). *Abbreviations*: PMMA, poly(methyl methacrylate); QD, quantum dot (From Ref. [38]. Reprinted with permission from AAAS.)

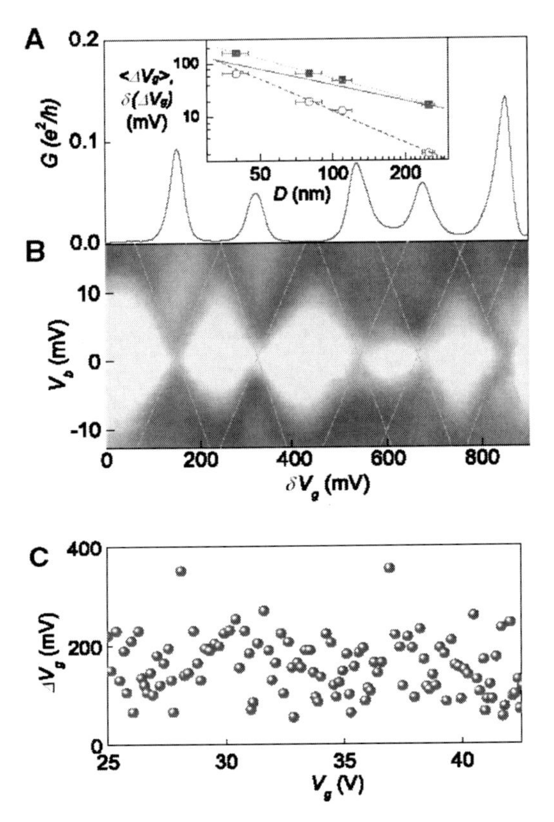

Figure 4.13 Effect of quantum confinement. CB peaks (A) and Coulomb diamonds (B) for a 40 nm QD (T = 4 K). Variations in peak spacing and the size of diamonds are clearly seen. The yellow-to-red scale in (B) corresponds to G_{diff} varying from 0 to 0.4 e^2/h. Two excited states (marked by additional lines) are feebly visible around $dV_g \approx 150$ and 850 mV and $V_b \approx 10$ mV. The smearing is caused by an increase in the transparency of quantum barriers at higher V_b. (C) Separation of nearest-neighbor peaks at zero V_b in the same device for a large interval of V_g (beyond this interval, CB became suppressed by high transparency of the barriers). Inset in (A): Log-log plot of the average peak spacing $\langle\Delta V_g\rangle$ (solid squares) and its standard deviation $\delta(\Delta V_g)$ (open circles) as a function of D. Linear ($\langle\Delta V_g\rangle \propto 1/D$, solid line) and quadratic ($\delta <\Delta V_g> \propto 1/D^2$, dashed) dependencies are plotted as guides to the eye. The dotted curve is the best fit for the average peak spacing: $\langle\Delta V_g\rangle$ $1/D^\gamma$ yielding $\gamma \approx 1.25$ (From Ref. [38]. Reprinted with permission from AAAS).

4.5 Fullerenes

Fullerene, an allotropic form of carbon, made up of spherical molecules formed from pentagonal and hexagonal rings, was first discovered in 1985. Because fullerenes have spacious inner cavities, atoms and clusters can be encapsulated inside the fullerene cages to form endohedral fullerenes. In particular, the unique structural and electronic properties of endohedral metallofullerenes (EMFs), where metal atoms are encapsulated within the fullerene, have attracted wide interest [39].

Potential electronic applications have rather focused on fullerenes in composites structures, such as adding them to CNTs. Figure 4.14 shows schematic models of the potential structure, whilst Fig. 4.15 shows transmission electron micrographs of actual EMF-filled CNTs. The encapsulated fullerenes can rotate freely in the space of a (10, 10) tube at room temperature, and the rotation of fullerenes will affect C60@(10, 10) peapod electronic properties significantly; generally, orientational disorder will remove the sharp features of the average density of states (DOS). However, the rotation of fullerenes cannot induce a metal–insulator transition. Unlike the multicarrier metallic C60@(10, 10) peapod, the C60@(17, 0) peapod is a semiconductor, and the effects of the encapsulated fullerenes on tube valence bands and conduction bands are asymmetrical.

CNTs doped with fullerenes inside nanotubes (so-called peapods) are interesting materials for novel CNT FET channels. Transport properties of various peapods such as C60-, Gd@C82-, and Ti2@C92 peapods have been studied by measuring FET I–V characteristics. EMF peapod FETs exhibited ambipolar behavior, with both p- and n-type characteristics achieved by changing the gate voltage, whereas C60-peapod FETs show unipolar p-type characteristics similar to FETs of SWCNTs. This difference can be explained in terms of a bandgap narrowing of the single-walled nanotube due to the incorporation of EMF. The bandgap narrowing was large in the peapods of EMF, where more electrons are transferred from encapsulated metal atoms to the fullerene cages.

The entrapped fullerene molecules are capable of modifying the electronic structure of the host tube. Hence, it is anticipated that the encapsulation of fullerene molecules can play a role in bandgap engineering in CNTs and hence that peapods may generate conceptually novel molecular devices.

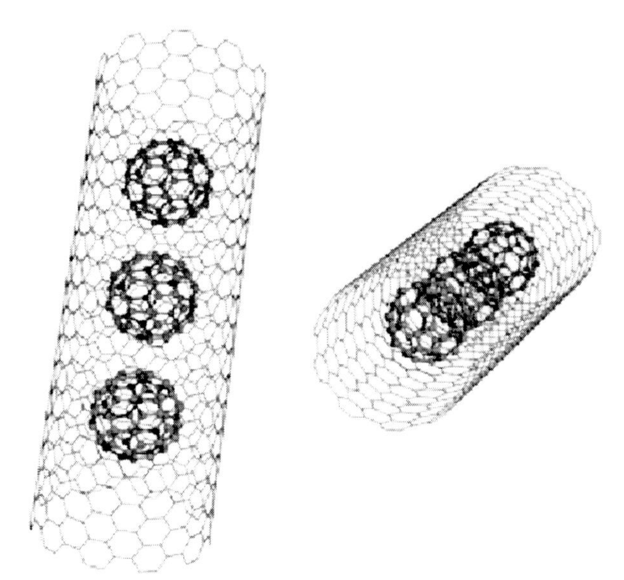

Figure 4.14 Schematics of the C60 fullerene peapods from two viewpoints (Reprinted with permission from Ref. [40]. Copyright Â© 2012 John Wiley & Sons, Ltd.).

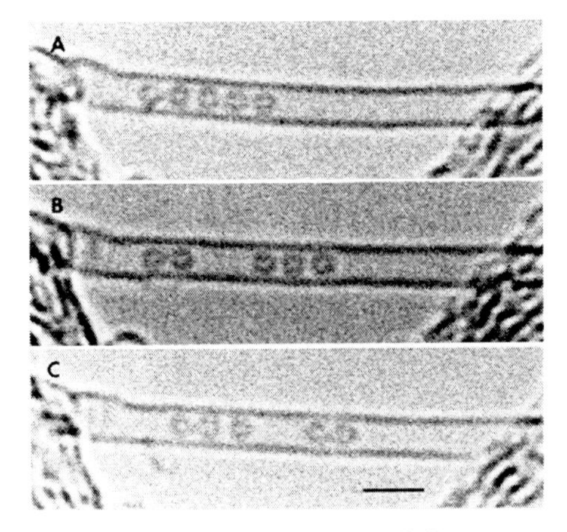

Figure 4.15 C60 molecules spontaneously diffusing inside a 1.4 nm diameter SWNT: (a) at 0 s; (b) after ~30 s; and (c) after ~60 s. From (a) to (b), a triplet has jumped 2 nm from its initial position towards the right-hand side of the image. From (b) to (c), a single C60 molecule has jumped back towards the left. Scale bar 2 nm (Reprinted from Ref. [41], Copyright (1999), with permission from Elsevier).

4.6 Conclusions

Carbon-based electronics is a vibrant research topic with great potential. Yet, despite the wealth of activity practical active devices based on diamond, DLC, CNTs, graphene, or fullerenes have yet to make a commercial impact on a large scale. Nevertheless, the potential for carbon-based electronics is great across a very broad range of applications.

References

1. www.e6cvd.com.
2. Johnson, E.O. (1965). Physical limitations on frequency and power parameters of transistors, *RCA Rev.*, **26**, pp. 163–177; Johnson, E.O. (1965). Physical limitations on frequency and power parameters of transistors, *IRE Int. Convent. Rec.*, **13**, pp. 27–34.
3. Keyes, R.W. (1972). Figure of merit for semiconductors for high-speed switches, *Proc. IEEE*, pp. 225–225.
4. Isberg, J. (2010). "Diamond electronic devices," in *Wide Bandgap Cubic Semiconductors: From Growth to Devices*, eds. Ferro, G., and Siffert, P. (American Institute of Physics, Strasbourg, France), pp. 123–128.
5. Denisenko, A., and Kohn, E. (2005). Diamond power devices. Concepts and limits, *Diamond Relat. Mater.*, **14**, pp. 491–498.
6. Twitchen, D.J., Whitehead, A.J., Coe, S.E., Isberg, J., Hammersberg, J., Wikstrom, T., and Johansson, E. (2004). High-voltage single-crystal diamond dioeds, *IEEE Trans. Electron Devices*, **51**, pp. 826–828.
7. Werner, M., Job, R., Zaitzev, A, Fahrner, W.R., Seifert, W., Johnston, C., and Chalker P.R. (1996). The relationship between resistivity and boron doping concentration of single and polycrystalline diamond, *Phys. Status Solidi (A)*, **154,** pp. 385–393.
8. Kajihara, S.A., Antonelli, A., Bernholc, J., and Car, R., (1991). Nitrogen and potential N-type dopants in diamond, *Phys. Rev. Lett.*, **66**, pp. 2010–2013.
9. Buckley-Golder, I.M., Bullough, R., Hayns, M.R., Willis, D.R., Piller, R.C., Blamires, N.G., Gard, G., and Stephen, J. (1991). Post-processing of diamond and diamond films: A review of some Harwell work, *Diamond Relat. Mater.*, **1**, pp. 43–50.
10. Braunstein, G., and Kalish, R. (1980). *Proceedings of the 2nd International Conference on Ion Beam Modification of Materials* (Albany, New York), pp. 691–697.

11. Prins, J.F. (1982). Biopolar transistor action in ion implanted diamond, *Appl. Phys. Lett.*, **41**, pp. 950–952.

12. See for example (2010). Proceedings of Diamond 2009, the 20th European conference on diamond, diamond-like materials, carbon nanotubes and nitrides, *Diamond Relat. Mater.*, **19**, pp. 351–672.

13. See for example (2010). Proceedings of the 3rd international conference on new diamond and nano carbons (NDNC) 2009, *Diamond Relat. Mater.*, **19**, pp. 107–272.

14. Hirama, K., Jingu, Y., Ichikawa, M., Umezawa, H., and Kawarada, H. (2009). DC and RF performance of diamond MISFET with alumina gate insulator, *Mater. Sci. Forum*, **600–603**, pp. 1349–1351.

15. Kasu, M., Ueda, K., Ye, H., Yamauchi, Y., Sasaki, S., and Makimoto, T. (2006). High RF output power for H-terminated diamond FETs, *Diamond Relat. Mater.*, **15**, pp. 783–786.

16. Ueda, K., Kasu, M., Yamauchi, Y., Makimoto, T., Schwitters, M., Twitchen, D.J., Scarsbrook, G.A., and Coe, S.E. (2006). Characterisation of high-quality polycrystalline diamond and its high FTE performance, *Diamond Relat. Mater.*, **15**, pp. 1954–1957.

17. Balmer, R.S., Friel, I., Woollard, S.M., Wort, C.J.H., Scarsbrook, G.A., Coe, S.E., El-Hajj, H., Kaiser, A., Denisenko, A., Kohn, E., and Isberg, J. (2008). Unlocking diamond's potential as an electronic material, *Phil. Trans. R. Soc. A: Math. Phys. Eng. Sci.*, **366**, pp. 251–265.

18. Kohn, E., and Denisenko, A. (2009). "Doped diamond electron devices," in *CVD Diamond for Electronic Devices and Sensors*, ed. Sussmann, R.S. (John Wiley & Sons, Chichester, U.K.), pp. 313–377.

19. Yamamoto, M., Watanabe, T., Hamada, M., Teraji, T., and Ito, T. (2005). Electrical properties of diamond p-i-p structures at high electric fields, *Appl. Surf. Sci.*, **244**, pp. 310–313.

20. Rashid, S.J., Tajani, A., Twitchen, D.J., Coulbeck, L., Udrea, F., Butler, T., Rupesinghe, N.L., Brezeanu, M., Isberg, J., Garraway, A., Dixon, M., Balmer, R.S., Chamund, D., Taylor, P., and Amaratunga, G. (2008). Numerical parameterization of chemical-vapor-deposited (CVD) single-crystal diamond for device simulation and analysis, *IEEE Trans. Electron Devices,* **55**, pp. 2744–2756.

21. Lettington, A.H. (1993). Thin film diamond, *Phil. Trans.: Phys. Sci. Eng.*, **342**, pp. 287–296.

22. Amaratunga, G.A.J., Chhowalla, M., Lim, K.G., Munindradasa, D.A.I., Pringle, S.D., Baxendale, M., Alexandrou, I., Kiely, C, J., and Keyse, B. (1998). Electronic properties of tetrahedral amorphous carbon (ta-C) films containing nanotube regions, *Carbon,* **36**, pp. 575- 579.

23. Iijima, S., and Ichihashi, T. (1993). Single-shell carbon nanotubes of 1-nm diameter, *Nat.*, **363**, pp. 603–605.

24. McEuen, P.L., Fuhrer, M.S., and Park, H. (2002). Single walled carbon nanotube electronics, *IEEE Trans. Nanotechnol.*, **1**, 78–85.

25. Choi, Y.M., Lee, D.-S., Czerw, R., Chiu, P.-W., Grobert, N., Terrones, M., Reyes-Reyes, M., Terrones, H., Charlier, J.-C., Ajayan, P.M., Roth, S., Carroll, D.L., and Park, Y.-W. (2003). Nonlinear behavior in the thermopower of doped carbon nanotubes due to strong, localized states, *Nano Lett.*, **3**, pp. 839–842.

26. Martel, R., Schmidt, T., Shea, H.R., Hertel, T., and Avouris, Ph. (1998). Single- and multi-walled carbon nanotube field-effect transistors, *Appl. Phys. Lett.*, **73**, pp. 2447–2449.

27. Heinze, S., Tersoff J., Martel R., Derycke V., Appenzeller, J., and Avouris, P. (2002). Carbon nanotubes as Shottky barrier transistors, *Phys. Rev. Lett.*, **89**, pp. 106801-1–106801-4.

28. Wind, S.J., Appenzeller, J., Martel, R., Derycke, V., and Avouris, Ph. (2002). Vertical scaling of carbon nanotube field effect transistors using top gate electrodes, *Appl. Phys. Lett.*, **80,** pp. 3817–3819.

29. Zhihong, C., Farmer, D., Sheng, X., Gordon, R., Avouris, Ph., and Appenzeller, J. (2008). Externally assembled gate-all-around carbon nanotube field effect transistor, *IEEE Electron Device Lett.*, **29**, pp. 183–185.

30. Farmer, D.B., and Gordon, R.G. (2006). Atomic layer deposition on suspended single-walled carbon nanotubes via gas-phase noncovalent functionalization, *Nano Lett.*, **6**, pp. 699–703.

31. Cao, J., Wang, Q., and Dai, H. (2005). Electron transport in very clean, as-grown suspended carbon nanotubes, *Nat. Mater.*, **4**, pp. 745–9.

32. Sangwan, V.K., Ballarotto, V.W., Fuhrer, M.S., and Williams, E.D. (2008). Facile fabrication of suspended as grown carbon nanotube devices, *Appl. Phys. Lett.*, **93**, pp. 113112-1–113112-2.

33. Lin, Y.-M., Tsang, J.C., Freitag, M., and Avouris, Ph. (2007). Impact of oxide substrate on electrical and optical properties of carbon nanotube devices, *Nanotechnology*, **18**, pp. 295202-1–295202-6.

34. www.physics.umd.edu/condmat/mfuhrer

35. Mueller, T., Kinoshita, M., Steiner, M., Perebeinos, V., Bol, A.A., Farmer, D. B., and Avouris, Ph. (2009). Efficient narrow-band light emission from a single carbon nanotube p–n diode, *Nat. Nanotechnol.*, **5**, pp. 27–31.

36. Geim, A.K., and Novoselov, K.S. (2007). The rise of graphene, *Nat. Mat*, **6**, pp. 183–191.

37. Li, X., Wang, X., Zhang, L., Lee, S., and Dai, H. (2008). Chemically derived, ultrasmooth graphene nanoribbon semiconductors, *Science*, **319**, pp. 1229–1232.
38. Ponomarenko, L.A., Schedin, F., Katsnelson, M.I., Yang, R., Hill, E.W., Novoselov, K.S., and Geim, A.K. (2008). Chaotic Dirac billiard in graphene quantum dots, *Science,* **320**, pp. 356–358.
39. Chai, Y., Guo, T., Jin C., Haufler, R.E., Chibante, L.P.F., Fure, J., Wang, L., Alford, J.M., and Smalley, R.E. (1991). Fullerenes with metals inside, *J. Phys. Chem.*, **95**, pp. 7564–7568.
40. Simon, F. and Monthioux, M. (2011) Fullerenes inside carbon nanotubes: the peapods, in *Carbon Meta-Nanotubes: Synthesis, Properties and Applications* (ed., M. Monthioux), John Wiley & Sons, Ltd, Chichester, UK, pp. 273–321.
41. Brian W. Smith, Marc Monthioux, David E. Luzz (199) Carbon nanotube encapsulated fullerenes: a unique class of hybrid materials, *Chem. Phys. Lett.*, **315**, pp. 31–36.

Chapter 5

Nanoscopic Interfacial Water Layers on Nanocrystalline Diamond: From Biosensors to Nanomedicine

Andrei P. Sommer and Hans-Jörg Fecht
Institute of Micro and Nanomaterials, Ulm University, Albert-Einstein-Allee 47, 89081 Ulm, Germany
andrei.sommer@uni-ulm.de

5.1 Introduction

Ever so often, a new term pops up that represents an emerging scientific trend. Biotechnology (104,000,000 Google hits), biomedical engineering, genetic engineering, tissue engineering, gene therapy, supramolecular chemistry, combinatorial chemistry, high-throughput screening, high-resolution imaging, single-molecule spectroscopy, and stem cell therapy, all are some representative examples of innovative trends. Recently, "nanotechnology" (26,700,000 Google hits) has become a popular term, representing the main efforts in current science and technology. Nanotechnology, sometimes called nanoscience, is still far from being a fully mature technology. By convention, nanotechnology refers to submicroscopic

Carbon-based Nanomaterials and Hybrids: Synthesis, Properties, and Commercial Applications
Edited by Hans-Jörg Fecht, Kai Brühne, and Peter Gluche

ISBN 978-981-4316-85-9 (Hardcover), 978-981-4411-41-7 (eBook)
www.panstanford.com

structures such as objects and films on the scale of 1–100 nm and also to specific techniques that allow us to both image and manipulate such structures.

In contrast to the aforementioned trends, nanotechnology represents not just one specific discipline. Nanotechnology involves an unprecedented degree of multidisciplinary expertise both in theory and in technical equipment. As a consequence, nanotechnology offers unexpected solutions to myriads of disciplines, including practical applications ranging from the design of new materials for space science in general and space exploration in particular, for instance, robust self-sufficient biosensors operating under extreme conditions, to personal care products such as sun blockers. The most recent branch of the nanotechnology tree is nanomedicine (1,260,000 Google hits, tendency increasing), which, according to the National Institutes of Health (NIH) Nanomedicine Roadmap Initiative, refers to highly specific medical intervention at the molecular scale for diagnosis, prevention, and treatment of diseases. In the course of the present contribution we will span a bridge ranging from basic high-resolution imaging methods in nanotechnology to novel applications in nanomedicine.

Nanomedicine, like no other previous technology, has the potential to revolutionize both diagnostics and therapeutics. The term "nanomedicine" can be traced back to the late 1990s [1]. In 2006 nanomedicine comprised biomedical fields as different as **drug delivery** (nanoparticles developed to improve the bioavailability drugs), **drugs and therapy** (nanoparticles used in the treatment of diseases that according to their structure have unique medical effects and differ from traditional small-molecule drugs), **in vivo imaging** (nanoparticle-based contrast agents for magnetic resonance imaging [MRI] and ultrasound imaging with better contrast and favorable biodistribution), **in vitro diagnostics** (biosensor concepts based on nanotubes, cantilevers, or atomic force microscopy), and **biomaterials** (biocompatible nanomaterials with improved mechanical and chemical properties for implants).

Initially, the term "nanomedicine" was limited to nanoscale drug delivery systems and nanoparticles—now called nanovehicles. Drug delivery systems range from veritable nanosystems (e.g., drug–polymer conjugates and polymer micelles) to microparticles in the range of 100 μm. Both nano- and microscale systems have been very important in developing clinically useful drug delivery systems [2].

Nanoparticles can be chemically functionalized depending on the targeted functions. For instance, in cancer treatment and detection nanoparticles serve multiple functions, including chemotherapy, thermotherapy, and radiotherapy. In advanced methods nanoparticles are combined with chemotherapy and photodynamic therapy [3]. These few examples are sufficient to illustrate the unprecedented multidisciplinary challenge emerging from nanotechnology. Small discoveries in one field can instantly initiate discoveries in fields that are topically remote. With nanotechnology, Feynman's $1,000 invitation "think small" has to be extended to "and think big." The enormous technological progress and never-seen-before multidisciplinary diversity that small discoveries, that is, discoveries concerning objects at the nanoscale, have triggered is only logical: the smaller the dimension of an object, the higher the probability that it is an element common to many macrosystems.

The extraordinary speed of the nanotech revolution is owned to the fact that the advance is taking place simultaneously in a multitude of fields. When 150 years ago Michael Faraday reported on the preparation of colloidal gold in nanometer size and of thin films of gold—*Experimental Relations of Gold (and Other Metals) to Light* [4]—there was practically no immediate use for gold nanoparticles. Faraday's original report and the recent explosive application of gold nanoparticles in nanotechnology in general, and in nanomedicine in particular, illustrate the big picture before and after the nanotechnology revolution.

5.2 Nanoscopic Interfacial Water Layers

There exists no surface in nature that is not covered by nanoscopically thin films of water, even in high vacuum. Whereas in the macroworld the presence of such water films is normally totally neglected, there is increasing observational evidence for their importance in the nanoworld, especially in biological systems. In discussing nanoscopic interfacial water layers it is instructive to proceed systematically and to start with the simplest system, the air–water interface. The air–water interface covers more than 70% of the earth's surface and strongly affects atmospheric, aerosol, and environmental chemistry. According to recent data revealed by isotopic dilution spectroscopy [5] order is found only in the topmost monolayer at the interface where water molecules present a free OH. In going over to liquid–liquid interfaces the picture becomes more complicated. Precondition for the observation of interfacial phenomena in liquid–liquid interfaces is immiscibility, as practically realized between a pure nonpolar liquid in contact with water. Interfacial ordering in such a system can be observed indirectly, for instance, by the order imposed to a third system such as a polymer, forming crystalline nanoneedles [6]—results confirming our own experiments performed both in stationary and in dynamical interfaces [7]. In dynamical interfaces one component is moving relative to the other—a process that can be associated with electrostatic processes involving the transfer of electric charge [8], which in turn can induce additional ordering.

In passing to liquid–solid interfaces, the degree of complexity of the system again increases, even in the case of water. In contrast to the air–water interface, where the order is limited to one monolayer presenting asymmetric charge distribution, the solid–liquid interface is complicated by the presence of a nanoscopic interfacial water layer, which is structurally different from bulk water. The difference manifests itself primarily in an increase in molecular order reflected by an increase in density and viscosity, depending on the particular nature of the solid surface. The situation is further complicated by the circumstance that the nanoscopic interfacial water layers persist both in air (where under normal relative humidity conditions the thickness of the adsorbed water film easily exceeds that of genuine nanoscopic interfacial water layers) and under water. In practice this means the coexistence of at least four phases: one monolayer of ordered water at the air–liquid interface on top of a layer of normal bulk water covering a nanoscopic film of interfacial water masking the solid surface.

In view of the complexity of the situation it is clear that experiments that allow us to analyze the order of nanoscopic interfacial water layers are in no way trivial. Initially the structure of nanoscopic interfacial water layers was probed by X-ray and neutron solution scattering experiments, which coherently indicated an increase in density on hydrophilic surfaces compared to normal bulk water. Analysis by molecular dynamics simulation confirmed the increase in density on hydrophilic surfaces [9].

5.2.1 Nanoscopic Interfacial Water Layers and Material Surfaces

Due to limited space we restrict the discussion of the structural analysis of nanoscopic interfacial water layers to research performed by the classical tools of nanotechnology: near-field scanning optical analysis (NOA) via a shear-force mode near-field scanning optical microscope (NSOM), interfacial force microscopy (IFM), friction force microscopy (FFM), atomic force microscopy (AFM), and atomic force acoustic microscopy (AFAM).

Experiments performed under ambient conditions by NOA provided quantitative evidence that the molecular organization of nanoscopic interfacial water layers on top of a moderately hydrophobic polymer film is different from that of normal bulk water [10]. The difference in organization manifested itself in an instant depletion of the nanoscopic interfacial water layers in response to their irradiation with 670 nm laser light—a wavelength that is practically not absorbed by normal bulk water. On hydrophilic surfaces nanoscopic interfacial water layers present densities and viscosities that are significantly higher than those for normal bulk water. Elevated viscosities were measured by IFM and FFM, coherently documenting a massive viscosity increase of the interfacial water layers [11, 12]. Complementary data on the persistence of hydration layers on hydrophilic surfaces in air and immersed in water were provided by AFM measurements [13]. Further insight into the nature of the nanoscopic interfacial water layers was provided by AFAM experiments performed under ambient conditions. AFAM analysis of the response of nanoscopic interfacial water layers on hydrophilic surfaces to their irradiation with 670 nm laser light applied at intensities as low as 50 W m^{-2} indicated an instant drop in both viscosity and density [14]. On hydrophobic surfaces the response of the nanoscopic interfacial water layers to irradiation with 670 nm laser light was an instant increase in contact stiffness, corresponding to a partial depletion of the nanoscopic interfacial water layers acting as a damping medium between the AFAM tip and the substrate [14].

Whereas hydrophilic surfaces exposed to ambient air are known to be covered with nanoscopic interfacial water layers, the prevalence of similar surface-covering wetting layers on hydrophobic surfaces is a matter of dogmatic discussions. Because of the practical impact of the wetting layer problem, both in nanotechnology and in nanomedicine, we summarize here the result of recent AFM experiments in which James et al. [15] provide clear and convincing evidence that nanoscopic interfacial water layers are almost universally present, even on apparently completely dry self-assembled monolayer surfaces considered most hydrophobic by conventional methods such as contact goniometry, thereby confirming our earlier results on the subject [16]. Additional information on the nature of nanoscopic interfacial water layers masking hydrophobic surfaces in ambient air was provided by AFAM. Exploration of the nanoscopic interfacial water layers on hydrogen-terminated nanocrystalline diamond substrates revealed a remarkable bonding stability, as opposed to corresponding water layers on top of nonhydrogenated nanocrystalline diamond [17].

The new insight regarding the order of the water molecules constituting the nanoscopic interfacial water layers on surfaces has a number of theoretical and practical implications. Apparently, order is not only imposed to the nanoscopic interfacial water layers by the substrate underneath, but the ordered water layers themselves possess the potential to induce order to molecules. On the basis of this capacity we established an origin-of-life model explaining the facilitated assembly of primordial amino acids—landing on natural hydrogenated diamonds covered with nanoscopic interfacial water layers—to the first polymers [18]. The catalytic capacity of nanoscopic interfacial water layers, and its interplay with laser light, has been recently exploited in the production of extremely ordered body-centered cubic carbon nanocrystals from a metastable carbon phase [19].

5.3 Nanomedicine

From the aforementioned perspective it is now evident that nanoscopic interfacial water layers must play a fundamental role in the nanoworld, in general, and in biological systems, in particular, where they determine the intrinsic contact between surfaces. In biological systems, nanoscopic interfacial water layers are involved in mediating the dynamical contact between biomaterial and cells [20]. Specifically, they seem to perform the function of informational blueprints during events of first contact [21–23]. In biosystems nanoscopic interfacial water layers are, however, not only involved in extracellular processes such as cellular recognition and attachment. In the interior of cells they constitute a considerable fraction of the mobile phase, which is lining membranes and filling the interface between macromolecules and organelles—predominantly hydrophilic surfaces.

Recently, we demonstrated that intracellular nanoscopic interfacial water layers respond collectively to irradiation with 670 nm laser light by an instant expansion of cytosol, explicable on the basis of the model experiments described previously [14]. The proof that it is possible to periodically modulate the density and viscosity of the nanoscopic interfacial water layers in cells, parameters that are increased relative to normal bulk water, is indirect: Using nondestructive intensities and energy densities of laser light [24] we showed that intermittent laser stimulation results in a powerful bidirectional flow across the cell membrane, thereby forcing the cells to incorporate high amounts of a drug in a short time [25]. Figure 5.1 illustrates the function of the facilitated drug uptake by laser-induced transmembrane convection (the light-cell pump), an innovative element in cancer chemotherapy. The potency of the convective drug uptake mechanism is documented in Fig. 5.2. Clearly, the validity of the convective transport is not restricted to

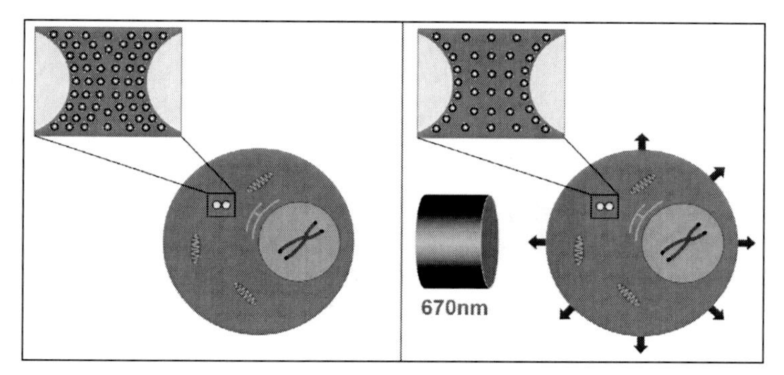

Figure 5.1 Eukaryotic cell with organelles (nucleus, endoplasmic reticulum, and mitochondria). Inset on the left: Interfacial water layers confined between macromolecules (e.g., two proteins). The cell on the right is irradiated with 670 nm laser light. On the nanoscale the irradiation reduces the density of the interfacial water layers in the space between the proteins (inset on the right). The related macroscale response in the extranuclear space is an instant efflux of cellular water (arrows).

unicellular systems. Extension of the cellular transport to tissues—as it is vital to increase the limited tumor penetration of anticancer drugs—is straightforward [26].

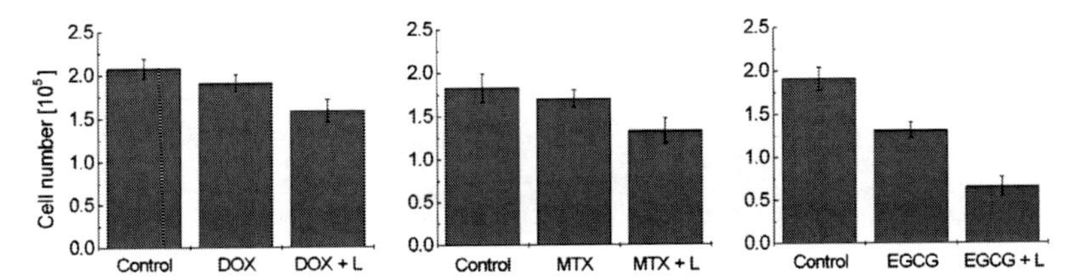

Figure 5.2 Intermittent 670 nm laser light modulates the density and viscosity of intracellular nanoscopic interfacial water layers and forces human cervical cancer cells (HeLa) to uptake anticancer compounds from their immediate environment. Columns from left to right: Control, cells supplemented with anticancer compounds (DOX, MTX, EGCG), and cells irradiated with laser light in the presence of an anticancer compound (DOX+L, MTX+L, ECCG+L). (Reprinted from Ref. [25], copyright 2010, with permission from Elsevier.) *Abbreviations*: DOX, doxorubicin; MTX, methotrexate; EGCG, epigallocatechin gallate.

Explicit evidence for the volume expansion of nanoscopic interfacial water layers in response to irradiation with 670 nm laser light, the root cause of convective transmembrane transport, followed from experiments performed at the Berlin Electron Storage Ring for Synchrotron Radiation (BESSY II), subsequently replicating the principle of the NOA-AFAM experiment [10, 14].Substitution of the high-resolution imaging machinery by soft X-rays provided unprecedented insight into details of the molecular structure of nanoscopic interfacial water layers under ambient conditions. Using moderately hydrogenated nanocrystalline diamond as a target substrate—to exclude artifacts due to possible contaminations of the ultrapure water used in the experiment [27, 28]—we demonstrated that nanoscopic interfacial water layers respond to 670 nm laser light by an instant volume expansion. The effect was reproduced on translucent polystyrene substrates (hydrophobic and hydrophilic) [29]. The discovery validates the mechanism thought to be causal for the process of transmembrane convection and provides a clear explanation of the previously observed spontaneous consolidation of nanoscopic interfacial water layers during dark phases [10, 14].

The soft X-ray experiment indicates that the process by which the nanoscopic interfacial water layers are spontaneously depleted in response to their exposure to 670 nm laser light is neither evaporation nor sublimation. The experimental setup, including the measuring chamber and the 670 nm laser, is shown in Fig. 5.3. The principle design and a corresponding photograph of the interior of the chamber are shown in Fig. 5.4. Figure 5.5 illustrates the analogy to the principle of the preceding AFAM experiment [14]. The results of the X-ray experiment are summarized in Fig. 5.6.

Figure 5.3 Combining 670 nm laser irradiation with soft X-rays obtained from a cyclotron radiation source to explore the molecular structure of interfacial water layers under ambient conditions.

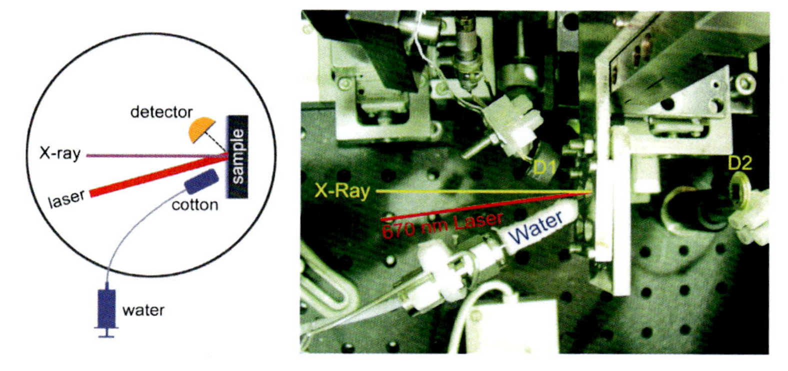

Figure 5.4 (Left) Self-explanatory image illustrates the principle of the soft X-ray experiment. (Right) The corresponding measuring chamber allows to modulate and probe nanoscopic interfacial water layers under ambient conditions.

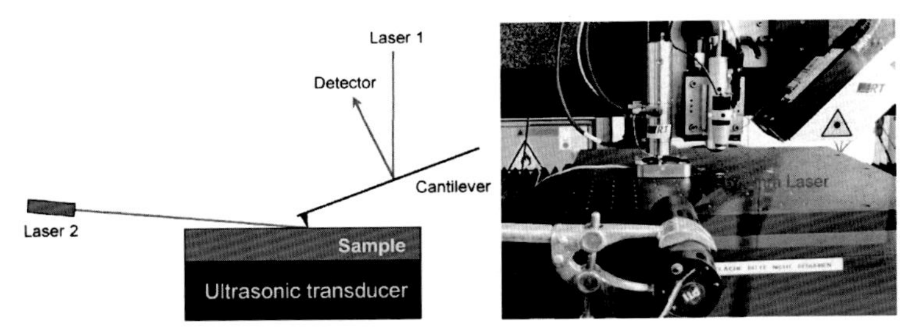

Figure 5.5 (Left) Principle of the AFAM experiment used to probe and modulate nanoscopic interfacial water layers on substrates with 670 nm laser light (laser 2). (Right) The corresponding experimental setup. The extension of the principle to the soft X-ray experiment is clear (cf. Fig. 5.4).

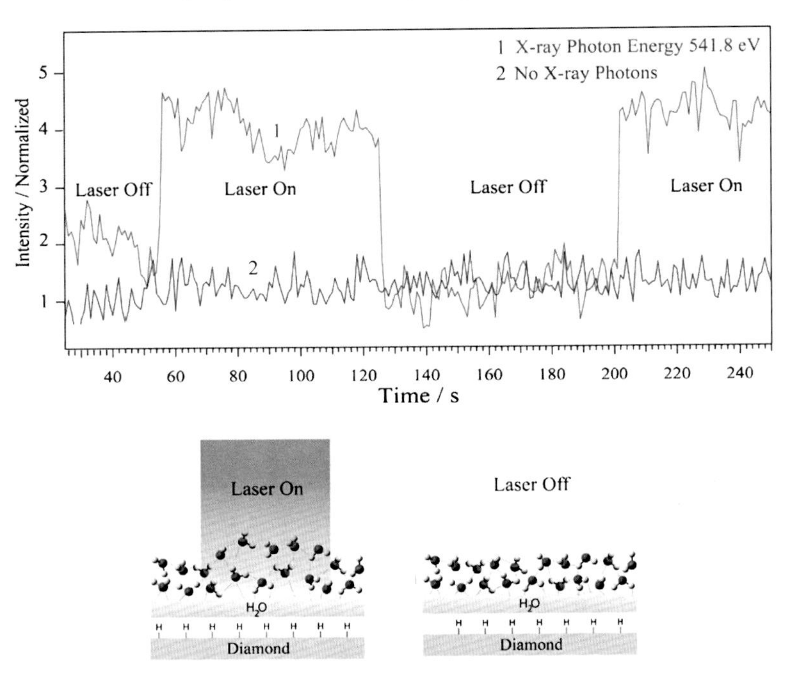

Figure 5.6 Strong, instant activation of nanoscopic interfacial water layers on a hydrogenated diamond surface by 670 nm laser light. The effect is interpreted as a breathing-like volume expansion of the topmost water layers and can be traced back to a collective excitation of water molecules. The effect is, to a large extent, independent of a specific substrate (Reprinted with permission from Ref. [29]. Copyright (2011) American Chemical Society).

5.4 Robust and Self-Sufficient Diamond-Based Biosensors

The importance of water as a precondition for life is best reflected via a Google search for the key words *water life Mars*, providing a total of 258,000,000 hits. Conventional hygrometers frequently employ plastic materials as humidity sensors. Because of the extreme radiation on Mars plastics have a very limited lifetime and are therefore not suitable for extended space missions. Actually, hygrometers designed for planetary exploration programs utilize an active polymer film as a humidity sensor, which changes capacitance as a function of relative humidity.

Temperatures at temperate and low Mars latitudes would permit, at least temporarily, the existence of liquid water on the planet's surface [30]. In this context, the development of robust and self-sufficient humidity sensor systems is an attractive challenge. Moreover, hygrometers designed for planetary exploration programs can also be used on Earth. Interestingly, hydrogen-terminated nanocrystalline diamond, including hydrogen-terminated natural diamond, responds to minimal variations in relative humidity, a finding that was reported by us earlier [18, 31]. It is based on the discovery that both synthetic and natural diamonds become electrically conductive by treatment with hydrogen plasma [32, 33] and exploit the (reproducible) phenomenon that conductivity decreases with increasing relative humidity [34]. Long-term stability of the hydrogen termination has been ascertained in the laboratory both for nanocrystalline [31] and for natural [18] diamonds. Specifically, there was no difference in electrical conductivity between new and six-year-old samples. Furthermore, the nanocrystalline diamond film and underlying substrate (silicon wafer) are individually durable. Finally, there were no signs of disruption between hydrogen-terminated diamond films and underlying silicon wafers.

However, according to recent laboratory experiments, diamonds may steadily lose atoms during exposure to ultraviolet (UV) light, even at very low fluences and even under incoherent illumination [35]. Whereas atom loss by UV irradiation is welcomed as a new technique for etching tiny features into diamond surfaces, it is clear that the effect might influence the quality of diamond-based hygrometers, in particular when operating for extended periods in conditions of extreme radiation. Figure 5.7 illustrates the principle for the construction of versatile humidity sensors based on hydrogen-terminated nanocrystalline diamond.

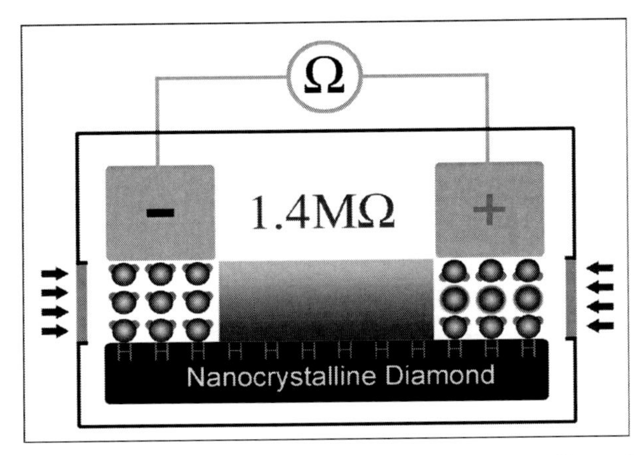

Figure 5.7 Principle of hydrogen-terminated nanocrystalline diamond–based relative humidity sensor (active sensor zone = electrode–substrate contact interface), which can simultaneously operate as a biosensor.

Considering that adhesion of microscopic/nanoscopic particles to the active sensor zone (i.e., the contact between electrodes and diamond) will change its characteristic conductivity pattern, on the one hand, and that the bioadhesive potential of organic particles such as proteins or microorganisms is superior to that of inorganic particles, on the other hand, we arrive at the design principle of a highly versatile biosensor, which can complementarily operate as a humidity sensor and a biosensor. Sand can be effectively excluded via suitable filters. To complete the modus operandi, we exploit the well-known capacity of hydrogen-terminated diamond to convert light into electrical energy. Noting that (partial) coverage of the nonactive sensor surface with organic material, for example, microorganisms, will affect the light-harvesting capacity but not the nanoscopic interfacial water layers in the active sensor zone (cf. Fig. 5.7), we are offered a realistic way to discriminate between variations in relative humidity and attachment of microorganisms, a new aspect demanding further research. Extension of the sensor functions by implementation of further components should be feasible, including the possibility to fingerprint biomarkers [36].

5.5 Nanoscopic Interfacial Water Layers on Hydrogen-Terminated Diamond: Model for Proton Transport in Cells

Hydrogen-terminated nanocrystalline diamond is not only interesting in technical applications, it is perhaps even more interesting as a platform to study transport processes involving protons in biosystems. Let us turn our attention to details of the implications of nanoscopic interfacial water layers in the effect of electrical conductivity, which is intrinsic to hydrogen-terminated diamond. From a rigorous analysis of the reciprocity between relative humidity and electrical conductivity, we arrived at the hypothesis that electrical conductivity is meditated by a Grotthuss-like proton hopping in the nanoscopic interfacial water layers established between electrodes (platinum) and the diamond surface [18] (cf. Fig. 5.7). Proton conductivity as a means to explain the unusual behavior of hydrogen-terminated diamond was proposed in 2001—a model based on the assumption that the effect is maintained by an exchange reaction at the diamond surface in the presence of moisture, recharging the diamond with protons [37]—yet without consideration of the water-structuring effect due to termination with hydrogen.

Proton transport belongs to the most basic processes in biological function and is not restricted to the plasma membrane. However, direct observation of the proton transport in biological systems is extremely difficult. During the past years there was an explosion of information on the many and varied roles of proton transport in cell function. Proton transport is involved in three broad areas of cell function: (a) maintenance and alteration of intracellular pH for initiation of specific cellular events, (b) generation of pH gradients in localized regions of the cell, including gradients involved in energy transduction, and (c) transepithelial ion transport. The processes involve one or more of several transport mechanisms [38]. Scientifically most exciting is proton hopping in biological systems—presumably the fastest-possible transport mechanism involving protons. It is instructive to quote from a very basic, almost visionary, paper of Szent-Györgyi: "Dielectric and conductivity measurements are reported for bovine serum albumin as a function of hydration. Strong evidence is found for the existence of mobile charges whose short- and long-range hopping motion strongly depends on the physical state of the protein-bound water. These charges are considered to be protons" [39]. Notably, proton hopping is a process characteristic for hydrophobic surfaces and as that relevant in biosystems with water in contact with hydrophobic proteins. Proton-hopping polarizability of water was recently predicted to be important in explaining rapid protein-folding rates, faster than predicted by simple kinetic theories. Proton hopping rather than dipolar reorientation is the cause of the high dielectric constant (high polarizability) of liquid water and ice. Protons move over large distances and create giant dipoles. This effect on the van der Waals forces has not been considered but may be expected to enhance the magnitude and/or range of the interaction. Since hydrophobic surfaces will only minimally restrict the motion of water molecules adjacent to them, it is expected that the enhanced polarizability of water at and between two hydrophobic surfaces could enhance the Lifshitz van der Waals–type attraction to a magnitude and/or range that is comparable to that between conductors, that is, by up to two orders of magnitude [40].

Whereas there is good evidence for proton hopping in biological systems, observability of the phenomenon is currently restricted to inanimate systems. In laboratory experiments proton hopping has been explored on hydrophobized surfaces, for instance, mica. Alternatively, fast proton transfer, including proton-coupled electron transfer, the mechanistic underpinning for radical transport, and catalysis in biology [41], prerequisites the prevalence of enzymes. In view of the fundamental importance of proton hopping in nanoscopic interfacial water layers within biological systems, on the one hand, and the lack of suitable model systems to experimentally study the process, on the other hand, we are realizing that by the molecular order imposed on water, hydrogen-terminated diamond is probably one of the best test and simulation platforms in the area of concern.

5.6 Conclusions

Using experimental methods as different as AFAM and soft X-ray absorption spectroscopy, we explored the structure of nanoscopic interfacial water layers on nanocrystalline hydrogenated diamond surfaces. The novel insight inspired innovative anticancer strategies and construction principles for a robust and self-sufficient biosensor, which recommends itself to be used to simultaneously discriminate between variations in relative humidity and attachment of microorganisms. The suitability of the nanocrystalline diamond-based biosensor for extended space exploration and planetary missions follows from the synergistic interplay of its properties (Fig. 5.8). However, the impact of the practical possibility to explore nanoscopic interfacial water layers on hydrogen-terminated nanocrystalline diamond surfaces exceeds these examples by far. Recently, Frauenfelder et al. demonstrated at the Los Alamos National Laboratory that the hydration shell of proteins plays a dominant role in controlling protein functions [42]. In other words, it is the nanoscopic interfacial water layer that controls protein functions and not vice versa. Computer simulations provided more clarity to this dominant role, regarding both the gluelike character and the elevated density of interfacial water layers masking hydrophilic proteins [43]. Therefore, the results of the aforementioned experiments hold the promise that 670 nm laser light can be used to selectively modulate the nanoscopic interfacial water layers constituting the hydration layer of proteins and thereby to nondestructively control protein functions in living cells—a fascinating possibility. Further research on hydrogen-terminated nanocrystalline diamond allowing us to fine-tune laser irradiation effects on nanoscopic interfacial water layers and test their biological impact is now necessary.

Figure 5.8 Films of hydrogenated nanocrystalline diamond on silicon wafers recommend themselves as ideal platforms for the construction of robust and self-sufficient biosensors capable of simultaneously operating as hygrometers and detectors of proteins and microorganisms and even of harvesting solar energy.

References

1. Wagner, V., Dullaart, A., Bock, A.K., and Zweck, A. (2006). The emerging nanomedicine landscape, *Nat. Biotechnol.*, **2**, pp. 1211–1217.

2. Park, K. (2007). Nanotechnology: what it can do for drug delivery, *J. Controlled Release,* **120**, pp. 1–3.

3. Khdair, A., Chen, D., Patil, Y., Ma, L., Dou, Q.P., Shekhar, M.P., and Panyam, J. (2010). Nanoparticle-mediated combination chemotherapy and photodynamic therapy overcomes tumor drug resistance, *J. Controlled Release,* **141**, pp. 137–144.

4. Faraday, M. (1857). The Bakerian lecture: experimental relations of gold (and other metals) to light, *Phil. Trans. R. Soc. Lond.*, **147**, pp. 145–181.

5. Stiopkin, I.V., Weeraman, C., Pieniazek, P.A., Shalhout, F.Y., Skinner, J.L., and Benderskii, A.V. (2011). Hydrogen bonding at the water surface revealed by isotopic dilution spectroscopy, *Nature,* **474**, pp. 192–195.

6. Nuraje, N., Su, K., Yang, N.L., and Matsui, H. (2008). Liquid/Liquid interfacial polymerization to grow single crystalline nanoneedles of various conducting polymers, *ACS Nano*, **2**, pp. 502–506.

7. Sommer, A. (1998). Flüssig/Flüssig-Grenzflächen: Strukturbildung und Stofftransport, PhD thesis (Marburg).

8. Sommer, A.P. (2006). Electrification vs crystallization: principles to monitor nanoaerosols in clouds, *Cryst. Growth Des.*, **6**, pp. 749–754.

9. Merzel, F., and Smith, J.C. (2002). Is the first hydration shell of lysozyme of higher density than bulk water? *Proc. Natl. Acad. Sci. U. S. A.*, **99**, pp. 5378–5383.

10. Sommer, A.P., and Franke, R.P. (2003). Modulating the profile of nanoscopic water films with low level laser light, *Nano Lett.*, **3**, pp. 19–20.

11. Goertz, M.P., Houston, J.E., and Zhu, X.Y. (2007). Hydrophilicity and the viscosity of interfacial water, *Langmuir*, **23**, pp. 5491–5497.

12. Jinesh, K.B., and Frenken, J.W.M. (2006). Capillary condensation in atomic scale friction: how water acts like a glue, *Phys. Rev. Lett.*, **96**, p. 166103.

13. Peng, C., Song, S., and Fort, T. (2006). Study of hydration layers near a hydrophilic surface in water through AFM imaging, *Surf. Interface Anal.*, **38**, pp. 975–980.

14. Sommer, A.P., Caron, A., and Fecht, H.J. (2008). Tuning nanoscopic water layers on hydrophobic and hydrophilic surfaces with laser light, *Langmuir*, **24**, pp. 635–636.
15. James, M., Darwish, T.A., Ciampi, S., Sylvester, S.O., Zhang, Z., Ng, A., Gooding, J.J., and Hanley, T.L. (2011). Nanoscale condensation of water on self-assembled monolayers, *Soft Matter*, **7**, pp. 5309–5318.
16. Sommer, A.P., and Pavláth, A.E. (2007). The subaquatic water layer, *Cryst. Growth Des.*, **7**, pp. 18–24.
17. Sommer, A.P., Zhu, D., Mester, A.R., Försterling, H.D., Gente, M., Caron, A., and Fecht, H.J. (2009). Interfacial water an exceptional biolubricant, *Cryst. Growth Des.*, **9**, pp. 3852–3854.
18. Sommer, A.P., Zhu, D., and Fecht, H.J. (2008). Genesis on diamonds, *Cryst. Growth Des.*, **8**, pp. 2628–2629.
19. Liu, P., Cui, H., and Yang, G.W. (2008). Synthesis of body-centered cubic carbon nanocrystals, *Cryst. Growth Des.*, **8**, pp. 581–586.
20. Sommer, A.P., Zhu, D., Franke, R.P., and Fecht, H.J. (2008). Biomimetics: learning from diamonds, *J. Mater. Res.*, **23**, pp. 3148–3152.
21. Sommer, A.P., Zhu, D., Wiora, M., and Fecht, H.J. (2008). The top of the biomimetic triangle, *J. Bionic Eng.*, **5**, pp. 91–94.
22. Sommer, A.P., Fuhrmann, R., and Franke, R.P. (2008). "Resistivity of cells under shear load," in *Metallic Biomaterial Interfaces*, eds. Breme J., Kirkpatrick, C.J., and Thull, R. (Wiley-VCH, Weinheim), pp. 242–245.
23. Sommer, A.P., Zhu, D., Scharnweber, T., and Fecht, H.J. (2010). On the social behaviour of cells, *J. Bionic Eng.*, **7**, pp. 1–5.
24. Sommer, A.P., Pinheiro, A.L.B., Mester, A.R., Franke, R.P., and Whelan, H.T. (2001). Biostimulatory windows in low-intensity laser activation: lasers, scanners, and NASA's light-emitting diode array system, *J. Clin. Laser Med. Surg.*, **19**, pp. 29–33.
25. Sommer, A.P., Zhu, D., and Scharnweber, T. (2010). Laser modulated transmembrane convection: implementation in cancer chemotherapy, *J. Controlled Release*, **48**, pp. 131–134.
26. Dumé, B. (2010). Red light speeds up drug uptake, *Nanotechweb.org.*, Nov. 17.
27. Willis, E., Rennie, G.K., Smart, C., and Pethica, B.A. (1969). Anomalous water, *Nature*, **222**, pp. 159–161.
28. Luck, W.A.P., and Ditter. W. (1970). Kann man anomales Wasser spektroskopisch in Kapillaren nachweisen? *Naturwissenschaften*, **57**, pp. 126.

29. Sommer, A.P., Hodeck, K.F., Zhu, D., Kothe, A., Lange, K.M., Fecht, H.J., and Aziz, E.F. (2011). Breathing volume into interfacial water with laser light, *J. Phys. Chem. Lett.*, **2**, pp. 562–565.

30. Möhlmann, D. (2010). The three types of liquid water in the surface of present Mars, *Int. J. Astrobiol.*, **9**, pp. 45–49.

31. Sommer, A.P., Zhu, D., and Brühne, K. (2007). Surface conductivity on hydrogen-terminated nanocrystalline diamond: implication of ordered water layers, *Cryst. Growth Des.*, **7**, pp. 2298–2301.

32. Landstrass, M.I., and Ravi, K.V. (1989). Resistivity of chemical vapor deposited diamond films, *Appl. Phys. Lett.*, **55**, pp. 975–977.

33. Landstrass, M.I., and Ravi, K.V. (1989). Hydrogen passivation of electrically active defects in diamond, *Appl. Phys. Lett.*, **55**, pp. 1391–1393.

34. Sommer, A.P., Zhu, D., and Försterling, H.D. (2008). Breathing conductivity into diamonds, *Science*, **February 28**.

35. Mildren, R.P., Downes, J.E., Brown, J.D., Johnston, B.F., Granados, E., Spence, D.J., Lehmann, A., Weston, L., and Bramble, A. (2011). Characteristics of 2-photon ultraviolet laser etching of diamond, *Opt. Mater. Express*, **1**, pp. 576–585.

36. Narayan, R.J., Boehm, R.D., and Sumant, A.V. (2011). Medical applications of diamond particles & surfaces, *Mater. Today*, **14**, pp. 154–163.

37. Koslowski, B., Strobel, S., and Ziemann, P. (2001). Are protons involved in the hydrogen-induced surface conductivity of diamond(001)? *Appl. Phys. A*, **72**, pp. 311–317.

38. Ives, H.E., and Rector, F.C., Jr. (1984). Proton transport and cell function, *J. Clin. Invest.*, **73**, pp. 285–290.

39. Gascoyne, P.R., Pethig, R., and Szent-Györgyi, A. (1981). Water structure-dependent charge transport in proteins, *Proc. Natl. Acad. Sci. U. S. A.*, **78**, pp. 261–265.

40. Hammer, M.U., Anderson, T.H., Chaimovich, A., Shell, M.S., and Israelachvili, J. (2010). The search for the hydrophobic force law, *Faraday Discuss.*, **146**, pp. 299–308.

41. Reece, S.Y., Hodgkiss, J.M., Stubbe, J., and Nocera, D.G. (2006). Proton-coupled electron transfer: the mechanistic underpinning for radical transport and catalysis in biology, *Philos. Trans. R. Soc. B*, **361**, pp. 1351–1364.

42. Frauenfelder, H., Chen, G., Berendzen, J., Fenimore, P.W., Jansson, H., McMahon, B.H., Stroe, I.R., Swenson, J., and Young, R.D. (2009). A

unified model of protein dynamics, *Proc. Natl. Acad. Sci. U. S. A.*, **106**, pp. 5129–5134.

43. Ahmad, M., Gu, W., Geyer, T., and Helms, V. (2011). Adhesive water networks facilitate binding of protein interfaces, *Nat. Commun.*, **2**, p. 261.

Chapter 6

Synthesis of Carbon Nanotubes and Their Relevant Properties

Aljoscha Roch, Esther Roch Talens, Beata Lehmann, Oliver Jost, and Andreas Leson
Fraunhofer Institute for Materials and Beam Technology, Winterbergstr. 28, 01277 Dresden, Germany
Aljoscha.Roch@iws.fraunhofer.de

6.1 Introduction

The first paper dealing with carbon nanotubes (CNTs) was published in 1952 by Radushkevich et al. in *Russian Magazine of Physical Chemistry* [1]. In 1976, further papers by Endo et al. followed, which showed images of CNTs taken by electron microscopy [2]. But methods to reliably synthesize nanotubes were still not known at that time. Initially, the interest of many groups of researchers had been aroused by the paper published by Iijima et al. in 1991 in *Nature*, because Iijiama found nanotubes in carbon black produced by electric arc evaporation and, thus, simultaneously a way to produce them [3]. These nanotubes found in 1991 were multiwalled carbon nanotubes (MWCNTs), and Iijiama further tried to fill them with

Carbon-based Nanomaterials and Hybrids: Synthesis, Properties, and Commercial Applications
Edited by Hans-Jörg Fecht, Kai Brühne, and Peter Gluche

ISBN 978-981-4316-85-9 (Hardcover), 978-981-4411-41-7 (eBook)
www.panstanford.com

transition metals. Performing these experiments he [4] and Bethune et al. [5] published, independently from each other, the discovery and synthesis of single-walled carbon nanotubes (SWCNTs) in 1993. Later, in 1996, Thess et al. introduced a synthesis based on a laser evaporation process that produces exclusively SWCNTs [6].

Hence, parallel to MWCNTs, SWCNTs have been a global subject of research activities, and efforts have been investigating their structure, physical properties, and synthesis. Particular properties of CNTs, as well as some applications, are stated in Chapter 8, which deals with the economical analysis of market opportunities for CNTs and nanodiamond by Werner et al.

6.2 Subdivision and Structure of Carbon Nanotubes

An SWCNT can be seen as a bended plane of graphene (a plane of carbon atoms with sp^2 orbitals) with caps similar to fullerenes [7]. A graphene sheet can be bent into an SWCNT in different configurations, depending on the angle $\Theta(n,m)$, which is shown in Fig. 6.1 for an SWCNT with the vector $\mathbf{C}(n,m)$ with $n = 5$ and $m = 2$.

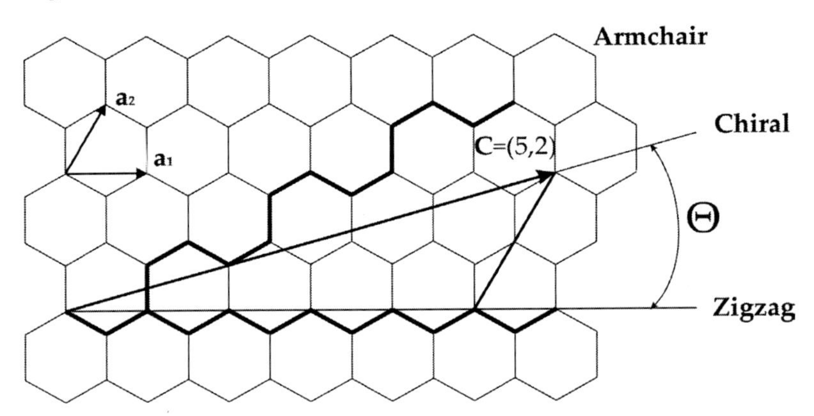

Figure 6.1 Graphene sheet with demonstration of a bending vector and angle.

The bending vector $\mathbf{C}(n,m)$ defines the direction for rolling the graphene plane. The vector is expressed by

$$\mathbf{C}(n,m) = n\,\mathbf{a_1} + m\,\mathbf{a_2}. \tag{6.1}$$

n and m are characteristic values belonging to each tube [8]. Thus, **C**(n,m) is limited to discrete values because of the impossibility to create arbitrary tube diameters by the factors m and n. The bending angle $\Theta(n,m)$ and the tube diameter d_t can be calculated knowing the characteristic factors n and m [7, 8]:

$$\Theta(n,m) = \tan^{-1}\left[\frac{m\sqrt{3}}{(m+2n)}\right], \tag{6.2}$$

$$d_t = \frac{|\mathbf{C}|}{\pi} = \frac{\sqrt{3}}{\pi}(m^2 + mn + n^2)^{0.5} a_{C-C}, \tag{6.3}$$

where $a_{C\text{-}C} = 1.44 \times 10^{-10}$ m is the distance between two carbon atoms. Depending on the bending angle $\Theta(n,m)$ the structures of nanotubes are differing, as are other physical properties. The possible arrangements are subdivided into three types shown in Fig. 6.2.

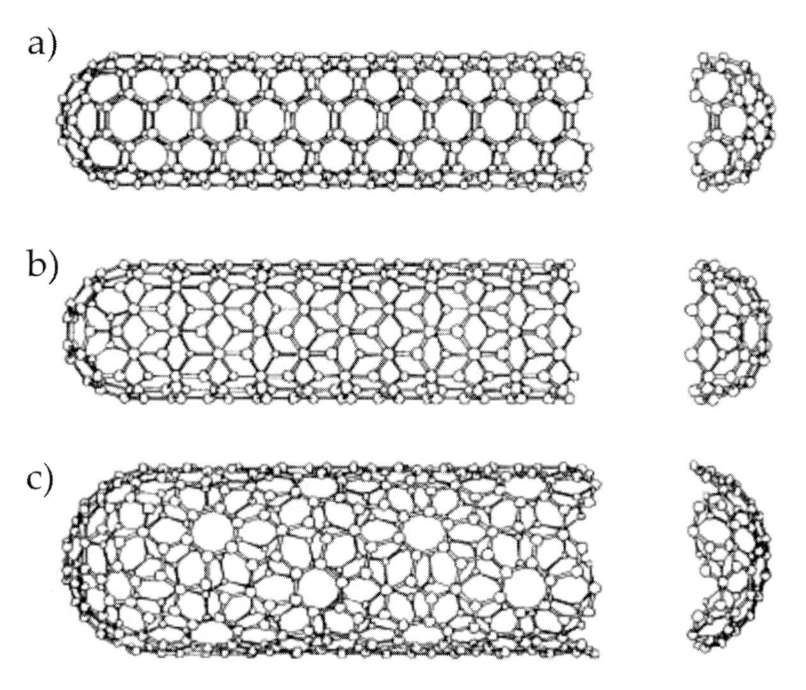

Figure 6.2 Illustration of different SWCNT arrangements (courtesy of R. Saito, G. Dresselhaus, and M. S. Dresselhaus, Ref. [8]).

The conditions for obtaining the different arrangements are:

(a) Armchair $\Theta(n,m) = 30°$

(b) Zigzag $\Theta(n,m) = 0°$

(c) Chiral $0° < |\Theta(n,m)| < 30°$

The diameter d_t of SWCNTs is around 0.4–5 nm [8, 9], and the length usually ranges between several nanometers up to some micrometers [8]. Yet, there have been grown nanotubes with a length up to some centimeters as well [10].

MWCNTs consist of several nanotubes with different diameters, concentrically attached to each other. Diameters of MWCNTs can reach dimensions of 100 nm. The length of MWCNTs is comparable to that of SWCNTs. The most inner tube of an MWCNT usually has a diameter of ≥5 nm, and each shell diameter is 0.7 nm larger, which corresponds to the 0.35 nm distance between graphene sheets in graphite (see Fig. 6.3) [7]. The diameter d_M of an MWCNT is given as

$$d_M = 2(r_i + 0.35 nm X), \tag{6.4}$$

where r_i is the radius of the most inner tube in an MWCNT and X represents the number of shells (Fig. 6.3).

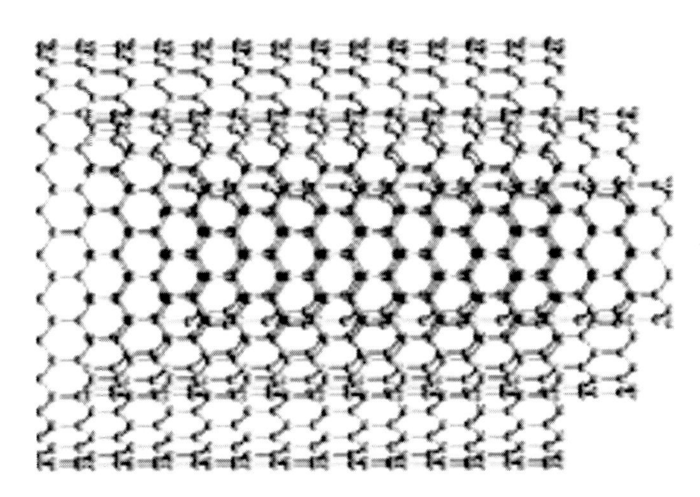

Figure 6.3 Schematic view of an MWCNT.

The ratios of volume V and mass G of an MWCNT compared to an SWCNT can be calculated if the tube lengths are equal and the tube radii r_S for the SWCNT and r_M for the MWCNT are known (Fig. 6.4).

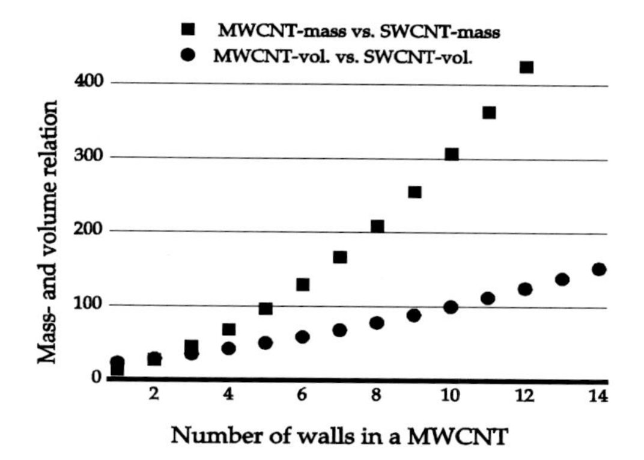

Figure 6.4 MWCNT and SWCNT (r_S = 0.6 nm, r_M = 2.5 nm) of the same length are compared by the criterion of the ratio of volume to mass.

If SWCNTs and MWCNTs have similar physical properties, the SWCNTs are seen as considerably advantageous (Fig. 6.4) because the ratio of "property per volume or mass" is remarkably larger.

6.3 Properties of SWCNTs

High potential has been predicted for SWCNTs because they demonstrate excellent physical properties in several scientific fields [7, 11, 12]. Their mechanical properties exceed those of most other materials in the range of some magnitudes. Thus, their ratio of tensile strength to density is much larger than in other materials (see Table 6.1).

Table 6.1 Comparison of mechanical properties of SWCNTs and other materials [12]

Material	Density in g/cm^3	Youngs modulus in MPa	Tensile strength in GPa	Tensile strength/Density (normalized)
SWCNT	1.4	542	65	462
MWCNT	1.8	400	2.7	15
Carbon fiber	1.6	152	2.1	13
Steel	7.6	207	0.7	1

The Young's modulus of SWCNTs is often stated to be 1.2 TPa [13, 14]. Further SWCNTs are ballistic conductors [12, 15] along their axis for electrons with extremely low ohmic resistance, which allows current densities j of 10^9 A/cm^2 [12, 16, 17]. An SWCNTs can transport up to 25 μA [15]. In metallic wires current densities j of $\sim 10^5$ A/cm^2 are allowed [17].

6.3.1 Electronic Band Structure of SWCNTs

The electronic band structure and electronic behavior of SWCNTs are defined by the bending angle $\Theta(n,m)$ and bending vector $\mathbf{C}(n,m)$ (Fig. 6.1), which can be calculated using the characteristic factors n and m of each tube, as described in section 6.2. The electronic band structure of SWCNTs $E_{1D}(\mathbf{k})$ is derived from the electronic band structure of graphene $E_{2D}(\mathbf{K})$ [7, 8]. For SWCNTs, the wave vector $\mathbf{k_1}$ pointing to the direction of the tube perimeter is restricted to discrete values s and depends on the tube diameter d_t [18]:

$$\left|\mathbf{k_1}\right| = \frac{2\pi}{\lambda} = \frac{2\pi}{|\mathbf{C}|} s = \frac{2s}{d_t}. \qquad (6.5)$$

The wave vector $\mathbf{k_2}$ in the axial direction of the tube can be continuous as it is in graphene, too.

$$\left|\mathbf{k_2}\right| = \text{arbitrary} \qquad (6.6)$$

Hence, the electronic band structure of SWCNTs $E_{1D}(\mathbf{k})$ is given by the equation

$$E_{1D}(\mathbf{k}) = E_{2D}(\mathbf{k}) \text{ with } \mathbf{k} = \mathbf{k_1} + \mathbf{k_2} = \mathbf{K}. \qquad (6.7)$$

The discrete values of $\mathbf{k_1}$ define line segments in the electronic band structure $E_{2D}(\mathbf{K})$ of graphene [8]. If a line segment is located in the metallic part of graphene's electronic band structure, the SWCNTs will have a metallic character as well [7]. This is true for all SWCNTs fulfilling the following condition, using the characteristic factors n and m:

$$3x = n - m \qquad (6.8)$$

In Eq. 6.8, x is an integer [8]. Thus, theoretically one-third of all possible SWCNTs are metallic, and two-thirds are semiconducting. The bandgap E_g of semiconducting SWCNTs (sc-SWCNT) ranges between 0.5 eV and 1 eV and is reciprocally proportional depending on the tube diameter d_t [15]:

$$E_{\mathrm{g}} \approx \frac{0.84\,\mathrm{eV}}{d_{\mathrm{t}}\,[nm]}. \tag{6.9}$$

This is also true for the single tubes of an MWCNT. However, because of the larger diameter d_{M} of MWCNTs, the real bandgap E_{g} normally is smaller than in SWCNTs.

6.3.2 Density of States and Optical Properties of SWCNTs

The electronic density of states $D_{1\mathrm{D}}(E)$ generally can be expressed in a one-dimensional (1D) system as [19]

$$D_{1\mathrm{D}}(E) = \frac{\sqrt{2m}}{\pi\hbar} \sum_{i}^{\infty} \frac{\Theta(E - E_{\mathrm{ii}})}{\sqrt{E - E_{\mathrm{ii}}}},$$

$$\text{with } \Theta(E - E_{\mathrm{ii}}) = \begin{cases} 0 & \text{for } E - E_{\mathrm{ii}} < 0 \\ 1 & \text{for } E - E_{\mathrm{ii}} \leq 0 \end{cases}. \tag{6.10}$$

For $E \rightarrow E_{\mathrm{ii}}$ that equation leads to singularities, called van Hove singularities.

For quasi-1D structures as SWCNTs, van Hove singularities appear in the density of states, as stated in Eq. 6.10. Between these singularities electronic transition and excitations can be excited, which are visible in absorbance spectra [20, 21].

Typical SWCNT-characterizing methods, such as optical and resonant Raman spectroscopy, are based on these dominant electronic excitations or excitons, respectively [8].

In Fig. 6.5, the upper part shows the absorbance spectrum of an SWCNT specimen. The so-called absorbance bands S_{11}, S_{22}, M_{11}, and S_{33}, resulting from the van Hove singularities, are shown as well. In the lower part in Fig. 6.5, the electronic band structure with the corresponding density of states $D_{1\mathrm{D}}(E)$ of metallic SWCNTs (m-SWCNTs) and sc-SWCNTs is shown.

For SWCNTs, the absorbance bands usually are identified as S_{ii} or M_{ii} [22, 25], where M (metallic) and S (semiconducting) represent the electronic properties of the nanotubes. The indices "11" and "22" state the type of transition (see Fig. 6.5).

Finally, the background, the π plasmon, typical for carbon in an sp^2-hybrid configuration is shown. It is resulting from the SWCNT but also from amorphous and graphitic carbon [23].

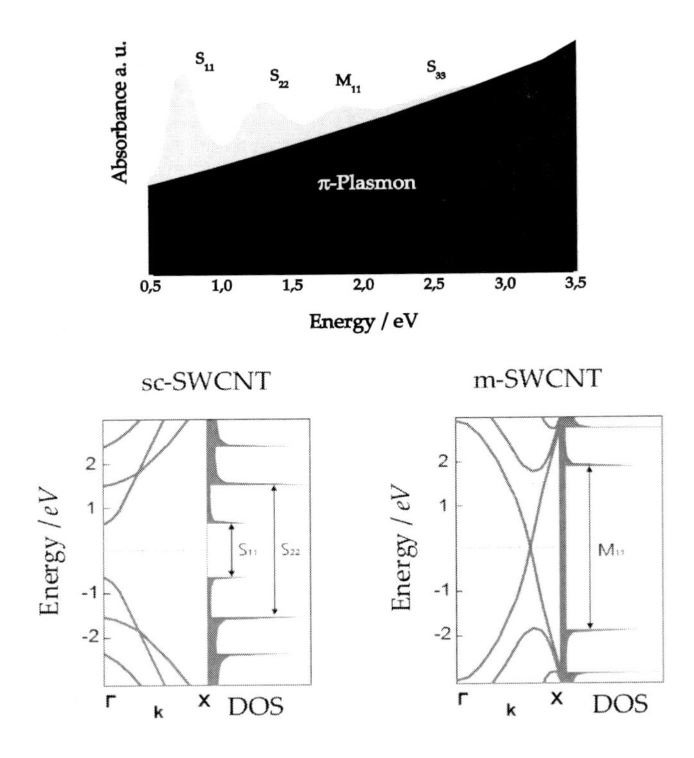

Figure 6.5 Absorbance spectra of an SWCNT specimen.

6.4 Characterization of SWCNTs

6.4.1 Optical Absorption Spectroscopy

Especially when using physical vapor deposition (PVD) synthesis, the produced "raw material" (usually a powder) contains, besides nanotubes, a lot of graphite, amorphous carbon, and several other carbon species. The percentage of these carbon impurities can be more than 50% [11, 24]. Ultraviolet-visible–near infrared (UV-Vis-NIR) spectroscopy is a rapid method to figure out the SWCNT yield of a specimen. The so-obtained absorbance spectra allow a quantification of the SWCNT percentage of the specimen [25]. Analysis of the spectra is done by comparing the absorbance bands with the background π plasmon. An example of the analysis is given in Fig. 6.6.

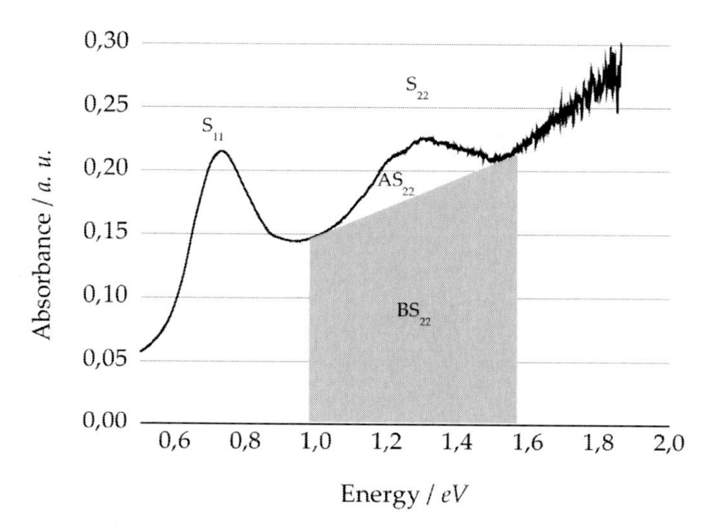

Figure 6.6 Diagram of the analysis by optical spectroscopy for an SWCNT specimen.

Characterization of the material is done by relating the surface area of the baseline-corrected S_{22} band with the background. Using Eq. 6.11, the RS_{22} value can be obtained.

$$RS_{22} = \frac{AS_{22}}{AS_{22} + BS_{22}} 100 \tag{6.11}$$

This value represents the proportional mass and volume ratio of SWCNTs in the specimen. The higher the RS_{22}, the more SWCNTs contained in the examined material.

Actually, the RS_{22} value is valid only for sc-SWCNTs and the RM_{11} value only for m-SWCNTs. That means that RM_{11} values obtained using Eq. 6.11 can be also quantitatively analyzed.

Unfortunately, RM_{11} and RS_{22} values of different specimens are not always comparable because of the different distribution of diameters, and so the quantitative analysis by optical absorption spectroscopy (OAS) of each SWCNT material has to be evaluated separately. Especially SWCNT material from different production processes can have different diameter distributions.

To get information on the SWCNT content in the raw material from the RS_{22} and RM_{11} values, the method has to be calibrated. This is done by purification by wet chemistry, eliminating amorphous carbon, graphite, and the catalyst. The values of RS_{22} and RM_{11} before and after purification are compared, as well as the mass loss; thus, a conversion factor can be calculated. When performing purification by wet chemistry, it is important to consider that the SWCNT surfaces could be functionalized by the process and, thus, absorbance bands might be suppressed. Hence, for optical analysis the RS_{22} value is usually used because it is less influenced by functionalization than RS_{11} [25, 26].

The average diameter d of an m-SWCNT and an sc-SWCNT can also be calculated from the absorbance bands using an empiric equation (Eqs. 6.12 and 6.13). The obtained values for the average diameter of SWCNTs are comparable to those calculated using Raman spectroscopy, as done by Milnera et al. [27]. However, calculating the average diameter using Raman spectroscopy is rather complicated; thus, optical spectroscopy seems more suitable. With Eq. 6.12, the average diameter d_{sc} of sc-SWCNTs can be calculated from the absorbance bands S_{11} and S_{22}.

$$S_{11} = \left[\frac{6}{5} (i+1) \frac{\gamma_0 \, a_{C-C}}{d_{sc}} \right]^{\frac{i+2}{i+1}} \tag{6.12}$$

S_{11} [eV] states the position of the band. Baseline correction has to be applied to the absorbance band, and then S_{11} can be obtained using Gauss approximation [28, 29]. γ_0 is a constant with the stated value of $\gamma_0 = 2.7$ eV [15]. The average diameter d_m of m-SWCNTs can be calculated with M_{11}:

$$M_{11} = \left[\frac{6}{5} (3+1) \frac{\gamma_0 \, a_{C-C}}{d_m} \right]^{\frac{5}{4}} \tag{6.13}$$

In Refs. [30, 21], further theoretic and empiric equations have been proposed for the calculation of the average diameter of SWCNTs.

For the characterization of an SWCNT powder, a flake of the SWCNT material is dispersed in isopropanol by auxiliary ultrasonication. The dispersion is then sprayed on a surface, and further analysis of that film results in absorbance spectra. The characteristic absorbance bands for nanotubes usually appear between 0.5 eV and 2.3 eV [20, 31]. Usually, bands on higher energy levels such as S_{33} or M_{22} are not considered.

6.4.2 Raman Spectroscopy

Resonant Raman spectroscopy is a useful method for detecting small amounts of SWCNTs by occurrence of the radial breathing modes (RBMs). Diameter distribution can be roughly estimated as well. Further estimation of the percentage of SWCNTs, defect density, and degree of functionalization is possible under certain conditions [29, 32–35]. Furthermore, specimens for Raman spectroscopy require low preparation, which is another advantage. So, an SWCNT isopropanol dispersion can be simply deposited on an aluminum foil.

In Raman spectra, the dominant peaks of SWCNTs are the RBMs between ~100 cm^{-1} and 350 cm^{-1} and the disorder modes (*D* and *D'* modes) at approximately ~1,350 cm^{-1} and ~2,600 cm^{-1}. Graphite shows a peak at 1,582 cm^{-1}, called the *G* mode, which can be explained by oscillation of the carbon atoms in the graphene sheet. Because of the bended graphene surface in SWCNTs, there the *G* mode is split into G^+ and G^- modes [29].

Besides the named dominant peaks there are minor peaks in the Raman spectrum of SWCNTs—according to the phonon dispersion relation of graphite (iTA, oTA, oTO, 2oTO, LA, iTOLA) (see Fig. 6.7) [36–38].

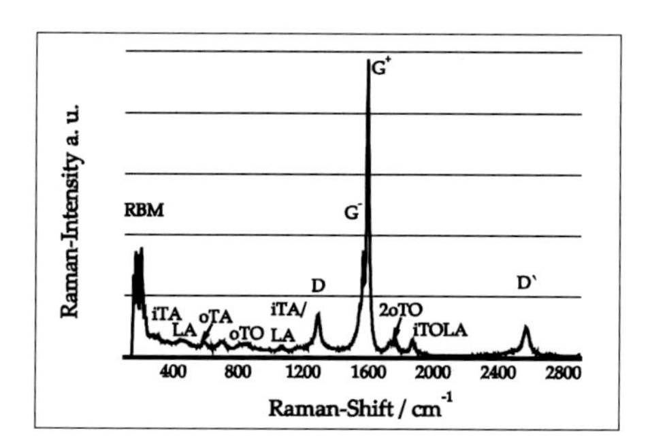

Figure 6.7 Raman spectrum of an SWCNT specimen.

6.4.2.1 Radial breathing mode

In RBMs, the carbon atoms of an SWCNT are oscillating in the radial direction. Because these oscillations are symmetrical, the tube seems to "breathe." This is caused by excited electrons switching from a van Hove singularity in the valence band to a van Hove singularity in the conduction band [8, 37]. This means that defined incident energies of the laser will cause a response to resonant Raman spectroscopy only at certain SWCNTs [34, 37]. Each RBM with a frequency of ω_{RBM} corresponds to an SWCNT of a certain diameter [34]:

$$\omega_{\mathrm{RBM}} = \frac{D}{d_{\mathrm{t}}} + B \tag{6.14}$$

For isolated SWCNTs on an oxidized silicon wafer, $D = 248$ cm^{-1} nm and $B = 0$ cm^{-1} [39], and for bundles of SWCNTs, $D = 234$ cm^{-1} nm and $B = 10$ cm^{-1} [29].

6.4.2.2 G^+ and G^- peaks

Contrary to other carbon materials, the Raman spectrum of SWCNTs shows a G peak split into a G^+ peak and a G^- peak. The G^+ peak and the G^- peak are composed of several subpeaks. The range in which different subpeaks can be anticipated is shown in Table 6.2 and Fig. 6.8.

Table 6.2 Split G peak for m- and sc-SWCNTs

	sc-SWCNT		m-SWCNT	
G^+ peak	1,592	[40]	~1,578–1,580	[29, 40]
G^- peak	1,553/1,569/1,608	[40]	1,530/1,544/1,548	[40]
	1,570	[29]	1,601/1,550	[29, 41]

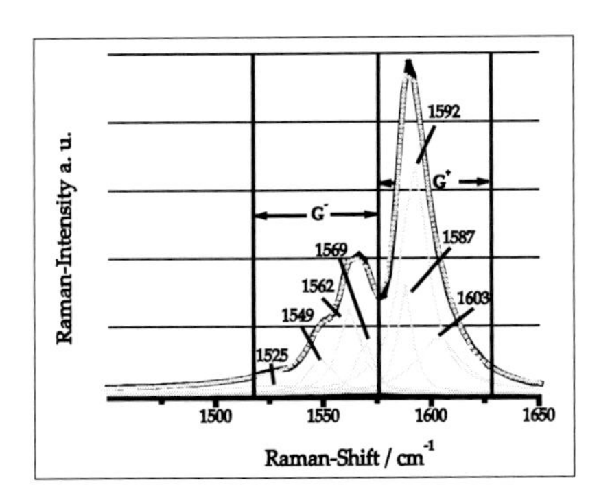

Figure 6.8 Raman fit of SWCNTs.

The split of the G peak in SWCNTs can be explained as follows: the G^+ peak is caused by longitudinal optical phonons along the axis of the tube, and the G^{-} peak is caused by transversal optical phonons along the perimeter of the tube [29]. The position of the G^{-} peak depends on the diameter of the tube and on its electronic properties [29].

Figure 6.9 shows the Raman spectrum of an SWCNT specimen obtained using a 785 nm and a 633 nm laser. The specimen's m-SWCNT reacted remarkably stronger to the 633 nm laser. If mainly m-SWCNTs are excited, coupling of the phonons and plasmons causes a Breit–Wigner–Fano (BWF) distribution of the G^{-} peak [40, 42].

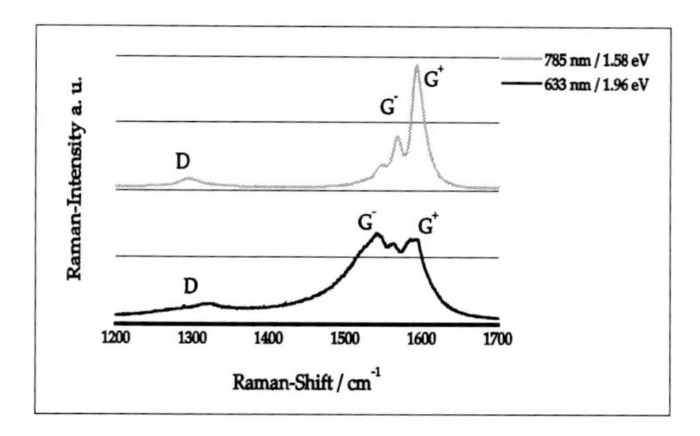

Figure 6.9 Raman shift at 785 nm and 633 nm.

6.4.2.3 *D* and *D′* peaks

The *D* peak appears in carbon with sp^2 orbitals if the symmetry is somehow disturbed (e.g., at the edges of a graphene sheet). Usually, it is found at a frequency of 1,350 cm^{-1} and explained with the occurrence of nanocrystalline carbon with sp^2 orbitals [43]. Besides, it can indicate amorphous carbon close to the SWCNTs or defects in the tubes [44]. Thus it is often assumed that the ratio of intensity between *D* and *G* peaks indicates the purity and quality of the SWCNT material [35, 45, 46].

However, Maultzsch et al. [44] approved that the chirality of nanotubes can influence the *D* peak as well. Therefore, the *D'* peak must be considered evaluating the defects, too. The *D'* peak appears at about 2,600 cm^{-1}. But, unlike the *D* mode, the *D'* mode cannot be explained with defects in the nanotubes [44, 47]. For a general analysis of specimens with different diameter distributions, the ratio of D'/D peak intensities shall be considered. For $D'/D \rightarrow \infty$ and $G/D \rightarrow \infty$, the defect density of SWCNT specimens is decreasing [44].

6.5 Synthesis of SWCNTs

SWCNTs can be produced using different processes. They are divided into chemical vapor deposition (CVD) and physical vapor deposition (PVD) [11, 15]. During PVD, both carbon and catalyst metals are evaporated and SWCNTs are synthesized in reactions during the cooling-down of the evaporated cloud. Usually, a laser or an electric arc is used to evaporate the carbon and the catalyst. Several different models have been developed to describe the synthesis, including the solid-liquid-solid model and the vapor-liquid-solid model [7, 15, 48, 49]. Furthermore, growth is described by the scooter or the sea urchin model [7]. However, the synthesis of SWCNTs has not been understood completely yet.

6.5.1 CVD for Synthesis of SWCNTs

In the CVD process, a carbon-containing gas—like methane (CH_4), ethene (C_2H_4), or carbon monoxide (CO)—is decomposed by high temperatures (600–1,400°C) and works as the carbon supply [50, 51]. Furthermore, a carrier gas, like nitrogen or argon, is used. Metallocenes like ferrocenes, which are decomposed by high temperatures as well, can be used as catalysts. Finally, the SWCNTs can grow on metallic droplets that originate from the decomposed metallocenes [50]. Another possibility to integrate the catalyst metal into the CVD process is the use of catalyst-coated substrates. These substrates are heated during the process until small metal droplets are generated (see Fig. 6.10).

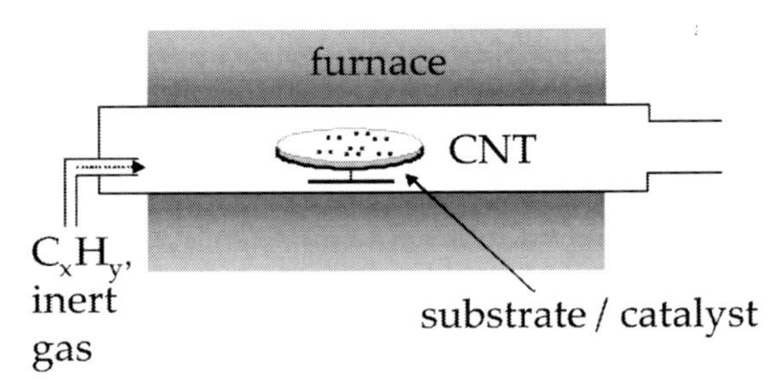

Figure 6.10 CVD synthesis of SWCNTs.

On top of the catalyst droplets, the SWCNTs will be synthesized. Metals often used as catalysts in CVD are cobalt (Co) and iron (Fe)

[52–54]. SWCNTs grown by CVD have diameters in the range of 0.7–4.0 nm [9, 24, 55] and can be rather long, in some cases up to some centimeters [10].

6.5.2 SWCNT Synthesis by Laser Evaporation

In this process carbon targets, doped with catalyst metals, are evaporated by short laser pulses, for example, by a neodymium:yttrium-aluminum-garnet (Nd:YAG) laser, and so SWCNTs are grown. The targets are evaporated in an oven with an inert gas atmosphere (Ar, He, or N_2) at temperatures of $T > 800°C$ (see Fig. 6.11).

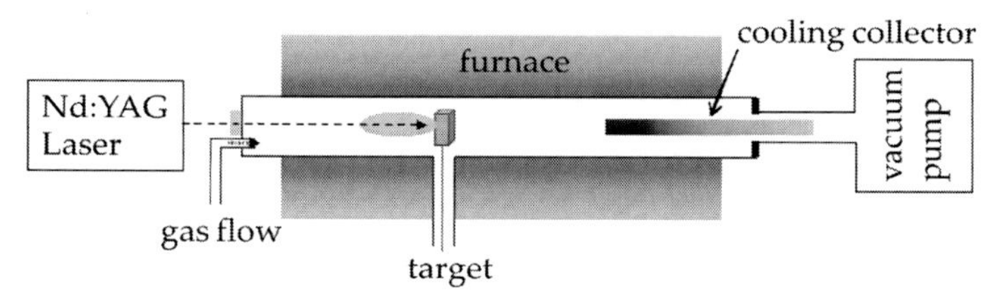

Figure 6.11 Laser synthesis of SWCNTs.

The energy of the laser is ~1 J, the pulse duration t_p is ~8 ns, and the pulse frequency f is ~20 Hz. The process can be influenced and optimized varying the named parameters [6, 56–59]. The pressure is usually between 660 mbar and 1,000 mbar [60]. The evaporation rate has magnitudes around ~20 mg/h. The evaporated material is carried by a flux of inert gas to a water-cooled collector, where it is deposited by thermophoresis. Normally, cobalt (Co) and nickel (Ni) are used as catalysts in the laser evaporation process [6, 49, 57]. The ratio of SWCNTs in the deposited material is ~20–90 wt% [11, 15, 57]. The diameter distribution can be influenced by the temperature, atmosphere gas of the oven, and catalyst [31, 61–64]. The average diameter d of the SWCNTs is between 1 nm and 2 nm [63]. Compared to electronic arc evaporation or CVD, laser evaporation is more laborious and economically disadvantageous [65].

6.5.3 SWCNT Synthesis by Continuous Arc Discharge

The technique originally used for the synthesis of SWCNTs is based on the continuous arc evaporation of a carbon target that is doped with catalytic metals. The current I is about 100 A, and usually, yttrium (Y) or nickel (Ni) is used as a catalyst [66, 67]. The evaporation is conducted in a water-cooled reactor filled with inert gases like argon (Ar) or helium (He) at gas pressure p of about 100–660 mbar [68] (see Fig. 6.12). The evaporated anode material is deposited on the chamber's walls, where it can be simply collected by chipping [68].

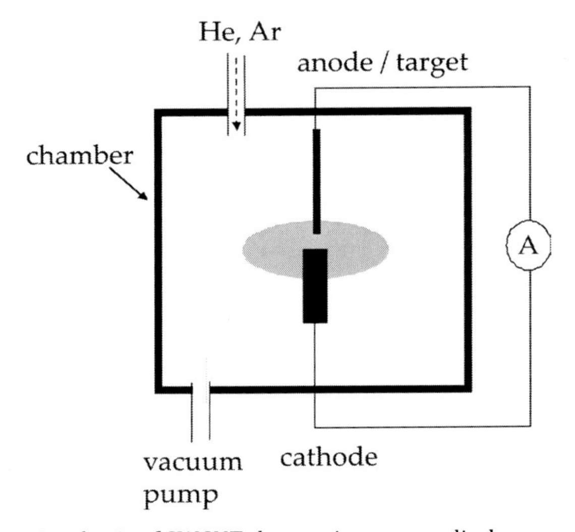

Figure 6.12 Synthesis of SWCNTs by continuous arc discharge.

The soot that is deposited on the chamber walls contains not only SWCNTs but also amorphous carbon, graphite, fullerenes, and MWCNTs. The average SWCNT content in the soot is about 10 wt% [24]. About 50% of the evaporated material is deposited on the cathode during the evaporation process. One advantage, however, is the relatively high evaporation rate achieved in a continuous arc of about 1 g/min [69]. The quite small diameter of the obtained SWCNTs is about 0.7–1.4 nm [15, 70].

6.5.4 SWCNT Synthesis by Pulsed Arc Discharge

The pulsed arc technique has been developed at the Fraunhofer Institute for Material and Beam Technology (FhG-IWS) in Germany [71]. This technique is based on an oven supported by pulsed arc evaporation (I = 400 A) of an anode. The reactor is made of silica glass, and the evaporation is carried out at a furnace temperature of 1,273 K (see Fig. 6.13). A stream of an inert gase (Ar, He, or N_2) continuously flows through the reactor in which the synthesis actually occurs. The pressure p during the synthesis is ≤300 mbar. As well as in laser evaporation, the used metal catalysts are transition metals like Ni or Co, which have shown high efficiencies. Additionally, iron (Fe), molybdenum (Mo), or rhenium (Re) as a catalyst metal or rare earths oxides like terbium oxide (Tb_3O_4) can be beneficial for the synthesis [72]. The average diameter d is ~1.25 nm, and the diameter distribution d_t = 0.9–1.6 nm. The length of the SWCNTs is ~10 μm.

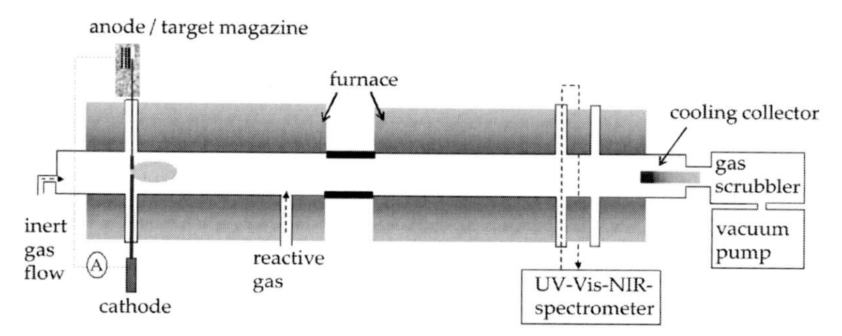

Figure 6.13 Pulsed arc process of SWCNT synthesis.

The pulsed arc technique offers several remarkable advantages compared to the already described processes like high manufacturing output and efficiency and is furthermore an economic technique. The production capacity is around 1 kg SWCNTs per day. The SWCNT yield in the produced soot is 40–80 wt%.

In Fig. 6.14, the evaporation techniques for SWCNTs are compared. The pulsed arc technique is remarkably advantageous compared to other evaporation processes.

6.5.5 Selective Synthesis of m- and sc-SWCNTs

Different approaches exist for the selective synthesis of m- and sc-SWCNTs.

Chiang et al. developed a CVD process for the selective synthesis of sc-SWCNTs. The SWCNT material produced by them contains up to 90% sc-SWCNTs. The catalyst mixture ($Fe_{83}Ni_{27}$) is denoted to be the reason for the selective synthesis of sc-SWCNTs [73].

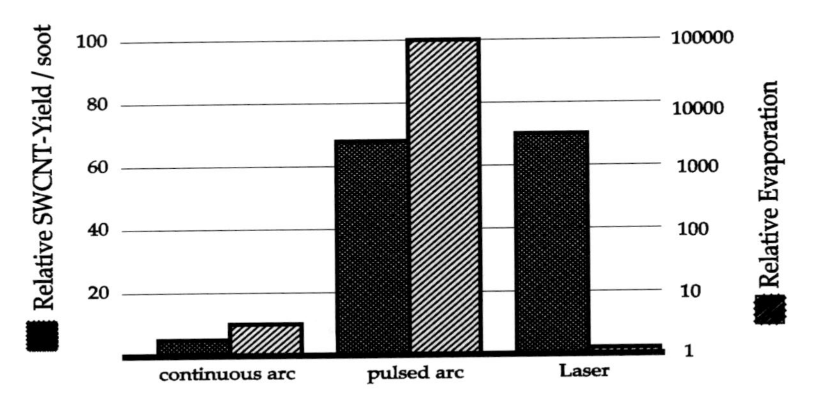

Figure 6.14 Comparison of evaporation techniques for SWCNT synthesis.

Li et al. used a plasma-enhanced CVD technique for the selective synthesis of up to 89% sc-SWCNTs. They suggested that the selective synthesis of sc-SWCNTs is caused by the lower formation energy and the higher chemical stability of the sc-SWCNTs in the Ar plasma. Metallic SWCNTs are much more sensitive to the etching effect of argon plasma [74].

Harutyunyan et al. showed that the gas atmosphere in CVD processes remarkably influences the structure of the synthesized sc-SWCNTs. They produced up to 91% sc-SWCNTs in a He atmosphere.

PVD processes have been proved to be capable of selective synthesis of CNTs, too. With the laser ablation technique Li et al. were able to produce up to 70% m-SWCNTs [75]. The arc technique was used by Rao et al. to synthesize up to 94% m-SWCNTs. They used a Ni/Y_2O_3 catalyst for SWCNT synthesis, and the process atmosphere was composed of He doped with iron pentacarbonyl, $Fe(CO)_5$ [76].

At the FhG-IWS the content of m-SWCNTs was found to be scalable between 20% and 70% depending on type of atmosphere and used catalyst. The mechanism of how selective synthesis works is not yet understood completely. However, it is assumed that for evaporation techniques with rather narrow diameter distribution, due to the few created chiralities, a shift of diameters has a stronger impact on the m-SWCNT/sc-SWCNT ratio. In Table 6.3, different suppliers of SWCNTs and the amount of m- and sc-SWCNTs in their material are shown.

Table 6.3 Comparison of different suppliers of SWCNTs and the relation of m- and sc-SWCNTs in the material [77]

Sample		Source	% sc-SWCNT	% m-SWCNT	% Standard error
HiPco[a]		Rice Univ.	62.9	37.1	0..5
CoMoCAT[b], standard grade		Univ. of Oklahoma	92.1	7.9	1.1
Laser ablation, low-temperature method		NASA-JSC	54.7	43.3	1.4
HiPco, metallic-enriched by DGU		Northwestern Univ.	2.6	97.4	0.4
CVD preferential growth		Current work	15.4	84.6	2.6
pulsed arc	m-enriched	FhG-IWS	34.8	65.2	3.0
	sc-enriched	FhG-IWS	72.1	27.9	3.0

[a] High-pressure carbon monoxide process for SWCNT synthesis [78].

[b] CVD process with Co and Mo catalysts [79].

Abbreviation: DGU, density-gradient ultracentrifugation.

6.6 Comparison of Different CNT Materials

6.6.1 Characterization of Purified Carbon Nanotubes by Raman and Transmission Electron Microscopy

It was found that due to differences in the synthesis techniques by several CNT suppliers, the materials show a large variety in their characteristics. There are, in general, differences in CNT diameter and length. As a consequence of the variation of these parameters, electrical conductivity as well as toxicity can be different. Furthermore, defect density as well as thermal and mechanical stability can differ, and unfortunately, it is often not stated what the relation of m- and sc-SWCNTs in the material is like.

The evaluation of the Raman spectra in Fig. 6.15 and other specifications of the CNT materials produced by different manufacturers are shown in Table 6.4.

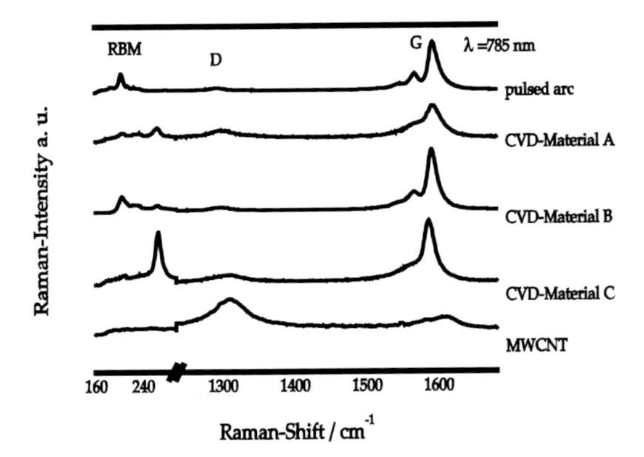

Figure 6.15 Raman spectra of different CNT materials produced by different CNT manufacturers.

The material produced by the pulsed arc technique and CVD material C seem to be similar. The main difference is observed in the fraction of m-SWCNTs, where the pulsed arc process (~50% m-SWCNTs) is favorable. The MWCNTs and the other CVD materials A and B contain shorter tubes and exhibit a much higher defect density, as can be seen from lower G/D and D'/D ratios.

Table 6.4 Comparison of different CNT materials synthesized by different producers and techniques.

	RS_{22}	Length in μm	Diameter in nm	m-SWCNTs in %	SWCNTs in wt%	G/D	D'/D
Pulsed arc	17	10	1.24	50	>90	12	2
CVD material A	3	>5	1–2	–	>90	6	0.4
CVD material B	1.5	A few	2	–	>70	3	0.1
CVD material C	7	3–50	1–1.3	20	>80	11	1.4
MWCNTs	–	5	>10	–	>95	<0.5	–

Figure 6.16 shows transmission electron micrographs of CNT materials. Especially for MWCNTs it can be observed that the tubes often are cracked and might have more defects. These pictures confirm the suggestions of the results made in the Raman evaluation. Therefore, the pulsed arc material and CVD material C are considered to have higher conductivity and shall be more suitable for electronic applications. The CNT material with a higher defect density might

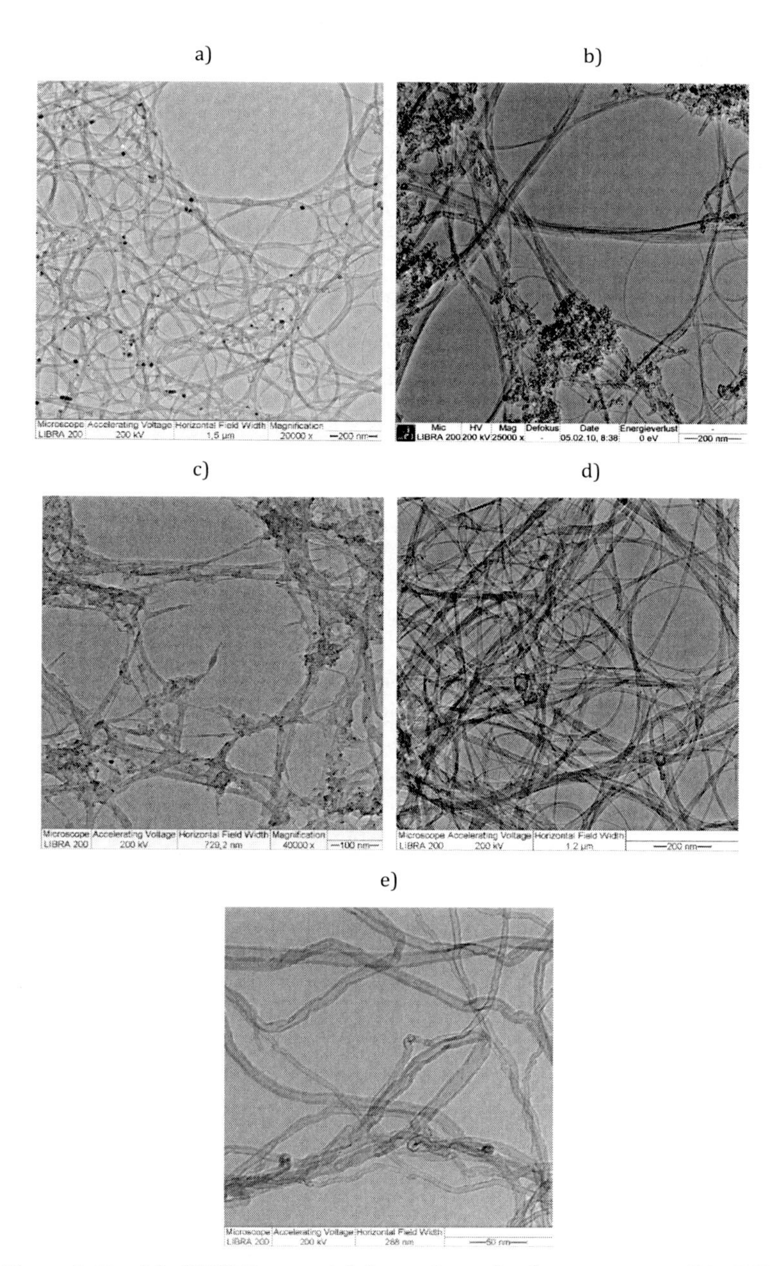

Figure 6.16 (a) SWCNT material from the pulsed arc process, (b) CVD material A, (c) CVD material B, (d) CVD material C, and (e) MWCNTs.

be better for mechanical applications to enhance the stiffness of materials. So, depending on the application an appropriate CNT material has to be chosen. In section 6.6.1.1, the use of CNT materials for transparent conducting films is discussed.

6.6.1.1 Characterization of transparent conductive films containing purified carbon nanotubes

Electrically conductive thin films of SWCNTs can be produced simply by spraying an SWCNT suspension on a substrate. When spraying a thin film, the homogeneity of the film is sensitively dependent on the degree of dispersion of the SWCNTs' agglomerates. There are aqueous and alcoholic SWCNT suspensions used for spraying. Using alcohol, the SWCNTs are dispersed in isopropanol under auxiliary ultrasonication and then deposited on a substrate with a spray pistol. The film's surface resistance can be measured by four-terminal sensing.

All materials introduced in section 6.6.1 can be used for the creation of conductive thin films. The material produced by the pulsed arc technique, however, contains ~50% m-SWCNTs, and Raman analysis shows low defect densities (see Table 6.4). It has been shown that thin films produced with that material are advantageous compared to films where CVD-grown MWCNTs or SWCNTs have been used, because electrical conductivity is higher at comparable transparencies (Fig. 6.17).

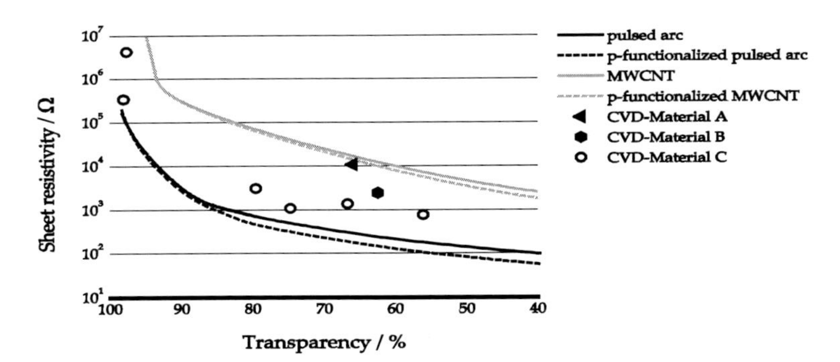

Figure 6.17 Transparency of electrically conductive surfaces made from different SWCNT materials.

It can be seen that the high percentage of m-SWCNTs and the low defect density in electric arc evaporation–grown material is advantageous for electrical applications. Further SWCNTs obtained from this process are rather long ($l \approx 10$ μm), which is beneficial for transparent conductive surfaces as well.

References

1. Radushkevich, L.V., and Lukyanovich V.M. (1952). O strukture ugleroda, obrazujucegosjapri termiceskom razlozenii okisi ugleroda na zeleznom kontakte, *Zurn. Fisic. Chim.*, **26**, pp. 88–95.
2. Oberlin, A., Endo, M., and Koyama, T. (1976). Filamentous growth of carbon through benzene decomposition, *J. Cryst. Growth.*, **32**(3), pp. 335–349.
3. Iijima, S. (1991). Helical microtubules of graphite carbon, *Nature*, **354**, pp. 56–58.
4. Ijima, S., and Ichihashi, T.S. (1993). Single-shell carbon nanotubes of 1-nm diameter, *Nature*, **363**, pp. 603–605.
5. Bethune, D.S., et al. (1993). Cobalt catalysed growth of carbon nanotubes with single-atomic-layer walls, *Nature*, **363**, pp. 605–607.
6. Thess, A., et al. (1996). Crystalline ropes of metallic carbon nanotubes, *Science*, **273**, pp. 483–487.
7. Dresselhaus, M.S., Dresselhaus, G., and Ph. Avouris (eds.). (2000). *Carbon Nanotubes* (Springer-Verlag, Berlin, Heidelberg, New York).
8. Saito, R., Dresselhaus, G., and Dresselhaus, M.S. (1998). *Physical Properties of Carbon Nanotubes* (Imperial College Press, London).
9. Lu, C., and Liu, J. (2006). Controlling the diameter of carbon nanotubes in chemical vapor deposition method by carbon feeding, *J. Phys. Chem. B*, **110**, pp. 20254–20257.
10. Zheng, L.X. et al. (2004). Ultralong single-wall carbon nanotubes. *Nat. Mater.*, **3**, pp. 673–676.
11. Daenen, M., de Fouw, R.D., Hamers, B., Janssen, P.G.A., Schouteden, K., and Veld, M.A.J. (2003). *The Wondrous World of Carbon Nanotubes* (Eindhoven University of Technology; interfacultary project; 93 pages).
12. Brand, L., Gierlings, M., Hoffknecht, A., Wagner, V., Zweck, A. (2009). *Kohlenstoff-Nanoröhren: Potenziale einer neuen Materialklasse für Deutschland, Zukünftige Technologien* (Consulting der VDI Technologiezentrum, Düsseldorf).

13. Krishnan, A., Dujardin, E., Ebbesen, T.W., Yianilos, P.N., and Treacy, M.M.J. (1998). Young's modulus of single-walled nanotubes, *Phys. Rev. B*, **58**, pp. 14013–14019.

14. Yu, M.-F. et al. (2000). Tensile loading of ropes of singlewall carbon nanotubes and their mechanical properties, *Phys. Rev. Lett.*, **84**, pp. 5552–5555.

15. O`Connell, M.J. (2006). *Carbon Nanotubes* (CRC Press, Taylor & Francis Group).

16. Yao, Z., Kane, C.L., and Dekker, C. (2000). High-field electrical transport in single-wall carbon nanotubes, *Phys. Rev. Lett.*, **84**, pp. 2941–2944.

17. Baughman, R.H., Zakhidov, A.A., de Heer, W.A. (2002). Carbon nanotubes—the route toward applications, *Science*, **297**, pp. 787–792.

18. Javey, A., Kong, J. (eds.). (2009). Chapter 1, "Band structure and electron transport physics of one-dimensional SWNTs," in *Carbon Nanotube Electronics. Series on Integrated Circuits and Systems* (Springer Science + Business Media), ISBN-10: 0-387-36833-7.

19. Waser, R. (2008). Nanotechnology Vol. 3: Information Technology I (Wiley-VCH Verlag).

20. Kataura, H., Kumazawa, Y., Maniwa, Y., Umezu, I., Suzuki, S., Ohtsuka, Y., and Achiba, Y. (1999). Optical properties of single walled carbon nanotubes, *Synth. Met.*, **103**, pp. 2555–2558.

21. Pedersen, T.G. (2004). Exciton effects in carbon nanotubes, *Carbon*, **42**, pp. 1007–1010.

22. Jeong, S.H., Kim, K.K., Jeong, S.J., An, K.H., Lee, S.H., Lee, Y.H. (2007). Optical absorption spectroscopy for determining carbon nanotube concentration in solution, *Synth. Met.*, **157**, pp. 570–574.

23. Sen, R., Rickard, S.M., Itkis, M.E., and Haddon, R.C. (2003). Controlled purification of single-walled carbon nanotube films by use of selective oxidation and near-IR spectroscopy, *Chem. Mater.*, **15**, pp. 4273–4279.

24. Krüger, A. (2007) *Neue Kohlenstoffmaterialien*, 1st ed. (Vieweg + Teubner, Deutschland), ISBN: 978-3-519-00510-0 (in German).

25. Itkis, M.E., Perea, D., Niyogi, S., Rickard, S., Hamon, M., Hu, H., Zhao, B., and Haddon, R.C. (2003). Purity evaluation of as-prepared single-walled carbon nanotube soot by use of solution-phase near-IR spectroscopy, *Nano Lett.*, **3**, pp. 309–314.

26. Itkis, M.E., et al. (2002). Spectroscopic study of the Fermi level electronic structure of single-walled carbon nanotubes, *Nano Lett.*, **2**, pp. 155–159.

27. Milnera, M., Kürti, J., Hulman, M., and Kuzmany, H. (2000). Periodic resonance excitation and intertube interaction from quasicontinuous distributed helicities in single-wall carbon nanotubes, *Phys. Rev. Lett.*, **84**, pp.1324–1327.

28. Liu, X., Pichler, T., Knupfer, M., Golden, M.S., Fink, J., Kataura, H., and Achiba, Y. (2002). Detailed analysis of the mean diameter and diameter distribution of single-wall carbon nanotubes from their optical response, *Phys. Rev. B*, **66**, pp. 045411-1–045411-8.

29. Jorio, A., Pimenta, M.A., Souza Filho, A.G., Saito, R., Dresselhaus, G., and Dresselhaus, M.S. (2003). Characterizing carbon nanotube samples with resonance Raman scattering, *New J. Phys.*, **5**, pp.139.1–139.17.

30. Jorio, A., Araujo, P.T., Doorn, St. K., Maruyama, S., Chacham, H., and Pimenta, M.A. (2006). The Kataura plot over broad energy and diameter ranges, *Phys. Status Solidi (B)*, **243**, pp. 3117–3121.

31. Ryabenko, A.G., Dorofeeva, T.V., and Zvereva, G.I. (2004). UV–VIS–NIR spectroscopy study of sensitivity of single-wall carbon nanotubes to chemical processing and Van-der-Waals SWNT/SWNT interaction. Verification of the SWNT content measurements by absorption spectroscopy, *Carbon*, **42**, pp. 1523–1535.

32. Graupner, R. (2007). Raman spectroscopy of covalently functionalized single-wall carbon nanotubes, *J. Raman Spectrosc.*, **38**, pp. 673–683.

33. Saito, R., and Kataura, H. (2001). Optical properties and Raman spectroscopy of carbon nanotubes, *Top. Appl. Phys.*, **80**, pp. 213–247.

34. Souza Filho, A.G., et al. (2003). Raman spectroscopy for probing chemically/physically induced phenomena in carbon nanotubes, *Nanotechnology*, **14**, pp. 2043–2061.

35. Saito, R., Grüneis, A, Samsonidze, Ge. G., Brar, V.W., Dresselhaus, G., Dresselhaus M.S., Jorio, A., Cancado, L.G., Fantini, C., Pimenta, M.A., and Souza Filho, A.G. (2003). Double resonance Raman spectroscopy of single-wall carbon nanotubes, *New J. Phys.*, **5**, pp. 157.1–157.15.

36. Saito, R., Jorio, A., Souza Filho, A.G., Dresselhaus, G., Dresselhaus, M.S., and Pimenta, M.A. (2002) Probing phonon dispersion relations of graphite by double resonance Raman scattering, *Phys. Rev. Lett.*, **88**, pp. 027401-1–027401-4.

37. Dresselhaus, M.S., Dresselhaus, G., Saito R., and Jorio, A. (2005). Raman spectroscopy of carbon nanotubes, *Phys. Rep.*, **409**, pp. 47–99.

38. Brar, V.W. et al. (2002). Second-order harmonic and combination modes in graphite, single-wall carbon nanotube bundles, and isolated

single-wall carbon nanotubes, *Phys. Rev. B*, **66**, pp. 155418-1–155418-10.

39. Jorio, A., Saito, R., Hafner, J.H., Lieber, C.M., Hunter, M., McClure, T., Dresselhaus, G., and Dresselhaus, M.S. (2001). Structural (n, m) determination of isolated single-wall carbon nanotubes by resonant Raman scattering, *Phys. Rev. Lett.*, **86**, pp. 1118–1121.
40. Brown, S.D.M., Jorio, A., and Corio P. (2001). Origin of the Breit-Wigner-Fano lineshape of the tangential G-band feature of metallic carbon nanotubes, *Phys. Rev. B*, **63**, pp. 155414-1–155414-8.
41. Brown, S.D.M., Corio, P., Marucci, A., Pimenta, M.A., Dresselhaus, M.S., and Dresselhaus, G. (2000). Second-order resonant Raman spectra of single-walled carbon nanotubes, *Phys Rev. B*, **61**, pp. 7734–7742.
42. Jiang, Ch., Kemp, K., Zhao, J., Schlecht, U., Kolb, U., Basche, Th., Burghard, M., and Mews, A. (2002). Strong enhancement of the Breit-Wigner-Fano Raman line in carbon nanotube bundles caused by plasmon band formation, *Phys. Rev. B*, **66**, pp. 161404-1–161404-4.
43. Ferrari, A.C., and Robertson, J. (2001). Resonant Raman spectroscopy of disordered, amorphous, and diamondlike carbon, *Phys. Rev. B*, **64**, pp. 075414-1–075414-13.
44. Maultzsch, J. (2004) *Vibrational Properties of Carbon Nanotubes and Graphite*, PhD thesis (Technische Universität, Berlin).
45. Costa, S., Borowiak-Palen, E., Kruszyńska, M., Bachmatiuk, A., Kaleńczuk, R.J. (2008). Characterization of carbon nanotubes by Raman spectroscopy, *Mater. Sci.—Poland*, **26**, pp. 433–441.
46. Kajiura, H., Tsutsui, S., Huang, H., and Murakami, Y., (2002). High-quality single-walled carbon nanotubes from arc-produced soot, *Chem. Phys. Lett.*, **364**, pp. 586–592.
47. Thomsen, Ch., Reich, St., and Maultzsch, J. (2004). Resonant Raman spectroscopy of nanotubes, *Phil. Trans. R. Soc. Lond. A*, **362**, pp. 2337–2359.
48. Gorbunov, A., Jost, O., Pompe, W., and Graff, A. (2002). Solid–liquid–solid growth mechanism of single-wall carbon nanotubes, *Carbon*, **40**, pp. 113–118.
49. Jost, O., Gorbunov, A.A., Möller, J., Pompe, W., Li, X., Georgi, P., Dunsch, L., Golden, M.S., and Fink, J. (2002). Rate-limiting processes in the formation of single-wall carbon nanotubes: pointing the way to the nanotube formation mechanism, *J. Phys. Chem. B*, 106, pp. 2875–2883.
50. Nikolaev, P. (2004). Gas-phase production of single-walled carbon nanotubes from carbon monoxide: a review of the HiPco process, *J. Nanosci. Nanotech.*, **4**, pp. 307–316.

51. Li, Y.-L. et al. (2004). Direct spinning of carbon nanotube fibers from chemical vapor deposition synthesis, *Science*, **304** (5668), pp. 276–278.

52. Park, J.B., Choi, G.S., Cho, Y.S., Hong, S.Y., and Kim, D. (2002). Characterization of Fe-catalyzed carbon nanotubes grown by thermal chemical vapor deposition, *J. Cryst. Growth*, **244**, pp. 211–217.

53. Inami, N., Mohamed, M.A., Shikoh, E., and Fujiwara, A. (2007). Synthesis-condition dependence of carbon nanotube growth by alcohol catalytic chemical vapor deposition method, *Sci. Technol. Adv. Mater.*, **8**, pp. 292–295.

54. Ishigami, N. Ago, H., Imamoto, K., Tsuji, M., Iakoubovskii, K., and Minami, N. (2008). Crystal plane dependent growth of aligned single-walled carbon nanotubes on sapphire, *J. Am. Chem. Soc.*, **130,** pp. 9918–9924.

55. Alexandrescu, R. et al. (2003). Synthesis of carbon nanotubes by CO_2 laser-assisted chemical vapour deposition, *Infrared Phys. Technol.*, **44**, pp. 4443–4450.

56. Guo, T., Nikolaev, P., Thess, A., Colbert, D.T., and Smalley, R.E. (1995). Catalytic growth of single-walled nanotubes by laser vaporization. *Chem. Phys. Lett.*, **243**, pp. 49–54.

57. Scott, C.D., Arepalli, S., Nikolaev, P., and Smalley, R.E. (2001). Growth mechanisms for single-wall carbon nanotubes in a laser-ablation process, *Appl Phys. A: Mater. Sci. Process.*, **72**, pp. 573–580.

58. Muñoz, E., Maser, W.K., Benito, A.M., de la Fuente, G.F., and Martínez, M.T. (1999). Single-walled carbon nanotubes produced by laser ablation under different inert atmospheres, *Synth. Met.*, **103**, pp. 2490–2491.

59. Arepalli S. (2004). Laser ablation process for single-walled carbon nanotube production, *J. Nanosci. Nanotechnol.*, **4**, pp. 317–325.

60. Gorbunov, A.A., Friedlein, R., Jost, O., Golden, M.S., Fink, J., and Pompe, W. (1999). Gas-dynamic consideration of the laser evaporation synthesis of single-wall carbon nanotubes, *Appl. Phys. A [Suppl.]*, **69**, pp. S593–S596.

61. Hinkov, I., et al. (2004). Effect of temperature on carbon nanotube diameter and bundle arrangement: microscopic and macroscopic analysis, *J. Appl. Phys.*, **95,** pp. 2029–2037.

62. Nishide, D., Kataura, H., Suzuki, S., Tsukagoshi, K., Aoyagi, Y., and Achiba, Y. (2003). High-yield production of single-wall carbon nanotubes in nitrogen gas, *Chem. Phys. Lett.*, **372**, pp. 45–50.

63. Jost, O., Gorbunov, A., Liu, X., Pompe, W., and Fink, J. (2004). Single-walled carbon nanotube diameter, *J. Nanosci. Nanotechnol.*, **4**, pp. 433–440.

64. Yudasaka, M., Ichihashi, T., and Iijima, S. (1998). Roles of laser light and heat in formation of single-wall carbon nanotubes by pulsed laser ablation of CxNiyCoy targets at high temperature, *J. Phys. Chem. B*, **102**, pp. 10201–10207.

65. Collins, P.G., and Avouris, P. (2000). Nanotubes for electronics, *Sci. Am.*, **283**, pp. 38–45.

66. Journet, C., Maser, W.K., Bernier, P., Loiseau, A., Lamy de la Chapelle, M., Lefrant, S., Deniard, P., Lee, R., and Fischer, J.E. (1997). Large scale production of single wall carbon nanotubes by the electric arc technique, *Nature*, **388**, pp. 756–758.

67. Mieno, T. (2006). Diffusion and reaction of carbon clusters in gas phase for production of carbon nanotubes, *New Diamond Frontier Carbon Technol.*, **16**, pp. 139–150.

68. Farhat, S., Hinkow, L., and Scott, C.D. (2004). Arc process parameters for single-walled carbon nanotube growth and production: experiments and modelling, *J. Nanosci. Nanotech.*, **4**, pp. 377–389.

69. Popov, Valentin N., and Lambin, P. (eds.) (2006). "Arc discharge and laser ablation synthesis of single-walled carbon nanotubes," in *Carbon Nanotubes: From Basic Research to Nanotechnology, Part I. Synthesis and Structural Characterization* (Springer-Verlag, New York), pp. 1–18, ISBN: 978-1-4020-4572-1.

70. Ebbesen, T.W., and Ajayan, P.M. (1992). Large-scale synthesis of carbon nanotubes, *Nature*, **358**, pp. 220–222.

71. Roch, A., Jost, O., Schultrich, B., and Beyer, E. (2007). High-yield synthesis of single-walled carbon nanotubes with a pulsed arc-discharge technique, *Phys. Status Solidi (B)*, **244**, pp. 3907–3910.

72. Roch, A., Märcz, M., Richter, U., Leson, A., Beyer, E., and Jost, O. (2009). Multi-component catalysts for the synthesis of SWCNT, *Phys. Status Solidi (B)*, **246**, pp. 2511–2513. Available at http://onlinelibrary.wiley.com/doi/10.1002/pssb.200982345/abstract.

73. Chiang, W.-H., Sakr, M., Gao, X.P.A., and Sankaran, R.M. (2009). Nanoengineering NixFe1-x catalyst for gas-phase, selective synthesis of semiconducting single-walled carbon nanotubes, *ASC Nano*, **3**, pp. 4023–4032.

74. Li, Y., Peng, S., Mann, D., Cao, J., Tu, R., Cho, K.J., and Dai, H. (2005). On the origin of preferential growth of semiconducting single-walled carbon nanotubes, *J. Phys. Chem. Lett. B*, **109**, pp. 6968–6971.

75. Li, Y., Mann, D., Rolani, M., Kim, W., Ural, A., Hung, S., Javey, A., Cao, J., Wang, D., Yenilmez, E., Wang, Q., Gibbons, J.F., Nishi, Y., and Dai, H.

(2004). Preferential growth of semiconducting single-walled carbon nanotubes by a plasma enhanced CVD method, *Nano Lett.*, **4**, pp. 317–321.

76. Voggu, R., Govindaraj, A., and Rao, C.N.R. (2009). A new method of obtaining high enrichment of metallic single-walled carbon nanotubes, *Mater. Sci.*, arXiv:0903.5359v1.

77. Naumov, A.V., Kuznetsov, O.A., Harutyunyan, A.R., Green, A.A., Hersam, M.C., Resasco, D.E., Nikolaev, P.N., and Weisman R.B. (2009). Quantifying the semiconducting fraction in single-walled carbon nanotube samples through comparative atomic force and photoluminescence microscopies, *Nano Lett.*, **9**, pp. 3203–3208.

78. Nikolaev, P. (2004). Gas-phase production of single-walled carbon nanotubes from carbon monoxide: a review of the HiPco process, *J. Nanosci. Nanotech.* **4**, pp. 307–316.

79. Lolli, G., Zhang, L., Balzano, L., Sakulchaicharoen, N., Tan, Y., and Resasco, D.E. (2006). Tailoring (*n,m*) structure of single-walled carbon nanotubes by modifying reaction conditions and the nature of the support of CoMo catalysts, *J. Phys. Chem. B*, **110**, pp. 2108–2115.

Chapter 7

Industrial Applications and Commercial Perspectives of Nanocrystalline Diamond

Matthias Wiora, Ralph Gretzschel, Stefan Strobel, and Peter Gluche
GFD, Gesellschaft für Diamantprodukte mbH, Lise-Meitner-Straße 13, 89081 Ulm, Germany
peter.gluche@gfd-diamond.com

Due to its outstanding and unique properties, diamond is used in a multiple of industrial fields, ranging from mechanical, thermal, and optical to electronic applications. Since natural diamond is rare, expensive, and unavailable in desired sizes and shapes, the applications are fairly limited. High-quality natural diamond (colorless, flawless, >0.01 carats) is evidently used for gemstones, whereas small stones or stones having impurities and large inclusions are used in industry, for example, in drilling, polishing, cutting, and sawing applications. The estimated entire world's production of mined natural diamonds in 2009 was about 160 million carats (equal to 32 tons), with 71 million carats thereof accounted for in industrial use [1]. The extreme hardness and high abrasion resistance of both natural and synthetic diamond make it a widely used material in industrial machining. Since synthetic diamonds cost from 15% to 40% less than naturally

Carbon-based Nanomaterials and Hybrids: Synthesis, Properties, and Commercial Applications
Edited by Hans-Jörg Fecht, Kai Brühne, and Peter Gluche

ISBN 978-981-4316-85-9 (Hardcover), 978-981-4411-41-7 (eBook)
www.panstanford.com

mined stones, it is not surprising that man-made diamonds account for 90% of the industrial market nowadays [2].

This chapter presents industrial applications and commercial perspectives of nanocrystalline diamond, with a focus on the utilization in mechanical watch movements. Diamond deposition and further processing concepts can be mostly adapted for application in the tooling industry and for any wear-resistant surface.

7.1 General Aspects and Applications

In any industrial application of chemical vapor deposition (CVD) diamond films, one has to consider the crucial choice of the substrate material, the operational environment (i.e., temperature), and the application function itself. The most critical substrate limitations are the following: (i) the high substrate deposition temperature (400–1,000°C), (ii) the film adhesion (difference in thermal expansion coefficient), and (iii) the material limitation to nonferrous materials (diffusion of carbon into the material). Despite these large restrictions, many successful products utilizing CVD diamond are available on the market nowadays, and a great number of highly potential new concepts are in development.

A well-established utilization of CVD diamond is in the tooling industry, especially the cutting, drilling, and milling of difficult-to-machine materials such as graphite, carbon-fiber-reinforced plastics, or plastic foils with metallic additives. Two examples of such applications are presented in Fig. 7.1.—a nanocrystalline diamond (NCD)-coated carbide tungsten milling tool for graphite machining (Fig. 7.1a) and an NCD-coated slit blade for plastic foil cutting (Fig. 7.1b), both significantly extending the lifetime in contrast to conventional steel or cemented carbide tools [3–5].

The optical and thermal properties of large-grained microcrystalline CVD diamond are ideal for optical windows (i.e., high-power laser windows, vacuum windows, microwave windows) or head spreaders (see Fig. 7.1d) [6]. Thick, free-standing diamond disks up to several hundreds of micrometers can be grown and mechanically postpolished to achieve smooth surfaces (Fig. 7.1c) [7]. Choosing growth conditions for large diamond crystals is essential to maintain optical transparency over a wide range, the low X-ray absorption, and the superior thermal conductivity.

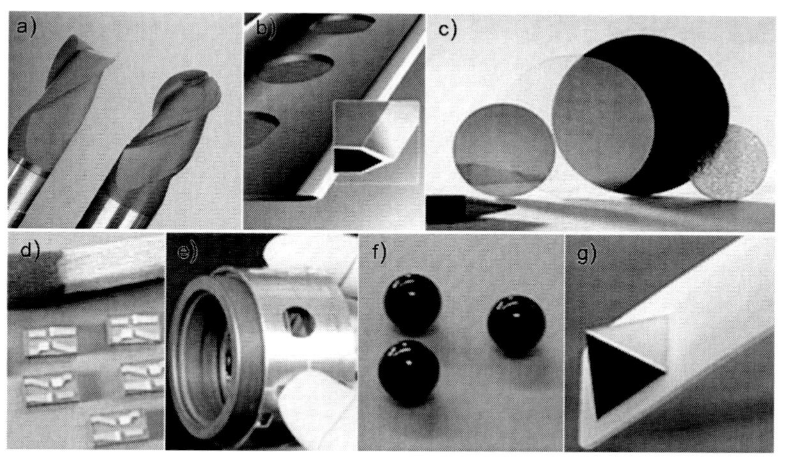

Figure 7.1 Industrial applications of CVD diamond. (a) NCD-coated carbide tungsten milling tool (Reprinted with permission of CemeCon AG). (b) NCD-coated carbide tungsten cutting blade, (c) diamond disks up to 15 cm in diameter and 2 mm in thickness with various microstructures (Reprinted from Ref. [7], with permission from Diamond Materials GmbH), (d) CVD microcrystalline heat spreaders with soldering and contact pads (Reprinted from Ref. [6], with permission from Diamond Materials GmbH), (e) NCD mechanical pump seal (Reprinted from Ref. [9] with permission of Advanced Diamond Technologies, Inc.), (f) NCD-coated tribological spheres, and (g) AFM probe entirely made out of NCD. (Reprinted from Auciello, O., and Sumant, A.V. (2010). Status review of the science and technology of ultrananocrystalline diamond (UNCD™) films and application to multifunctional devices, *Diamond Relat. Mater.*, **19**(7–9); with permission from Elsevier.)

The field of tribology particularly takes advantage of the high wear resistance and the low coefficient of friction of diamond. NCD with its as-grown smooth surface and high mechanical stability consequently is an excellent surface protection coating. The high shock and wear resistance becomes especially effective in applications such as bearings, pump seals (Fig. 7.1e), or tribological spheres (Fig. 7.1f), both enhancing the reliability and lifetime [8, 9]. Furthermore, the precise growth and microfabrication of NCD allows structures entirely made of diamond, such as diamond atomic force microscopy (AFM) probes produced in a monolithic process (Fig. 7.1g), exhibiting imaging performance comparable to standard probes but with over a 100 times less wear rate [10].

7.2 CVD Diamond in a Mechanical Watch Movement

The combination of high wear resistance and minimal friction of smooth diamond surfaces is ideal for application in micromechanical systems. Here, a special and highly interesting field is mechanical watch movements, where accuracy, reliability, and lifetime are crucial factors. Furthermore, diamond surfaces in a clockwork have the potential of lubrication-free watch movement and diminishing of all the disadvantages of oil, such as aging, accumulation of dirt, and, consequently, frequent service intervals.

Within a mechanical clockwork, the escapement mechanism is the most important part determining accuracy and energy consumption. Generally, escapement is responsible for counting the regulating organ's oscillations and supplying it with energy at the same time.

A mechanical watch is made of five essential parts, including the mainspring (power supply), the gear train, and the escapement mechanism. The Swiss lever escapement mechanism in turn consists of five components: a balance wheel, a hairspring, an anchor with two pallets, and an escapement wheel. The balance wheel oscillates periodically under the driving force from the escapement wheel through the pallet fork and restores the force of the hairspring. The impulse pin, which is a synthetic ruby on the balance wheel, sends and receives impulses from the fork to the balance wheel. The banking pin limits the rotation of the fork. The escapement wheel rotates periodically at a specific speed according to the frequency of the system.

Figure 7.2 illustrates the working mechanism of Swiss lever escapement and the different states in a full cycle [11–13]. There are a total of 10 main events (shocks, beat noises, and sliding) during this period (see numbers 1–10 in Fig. 7.2). In stage A, the balance wheel is turning counterclockwise with a tooth of the escapement wheel locked with the left pallet (entry pallet), and the lever is held in place by the left banking pin. Stage B shows the unlocking of the wheel. The impulse pin on the balance wheel enters and impacts the fork, giving the first shock (1). Then it pushes the pallet fork to release the tooth (2), and the tooth slides clockwise up the face of the pallet (3), providing an impulse to the balance wheel (stage C). This sliding motion (3) is especially critical, since for common materials careful lubrication is essential, otherwise the high friction between the tooth and the pallet leads to high wear, power losses, and, finally, failure of movement.

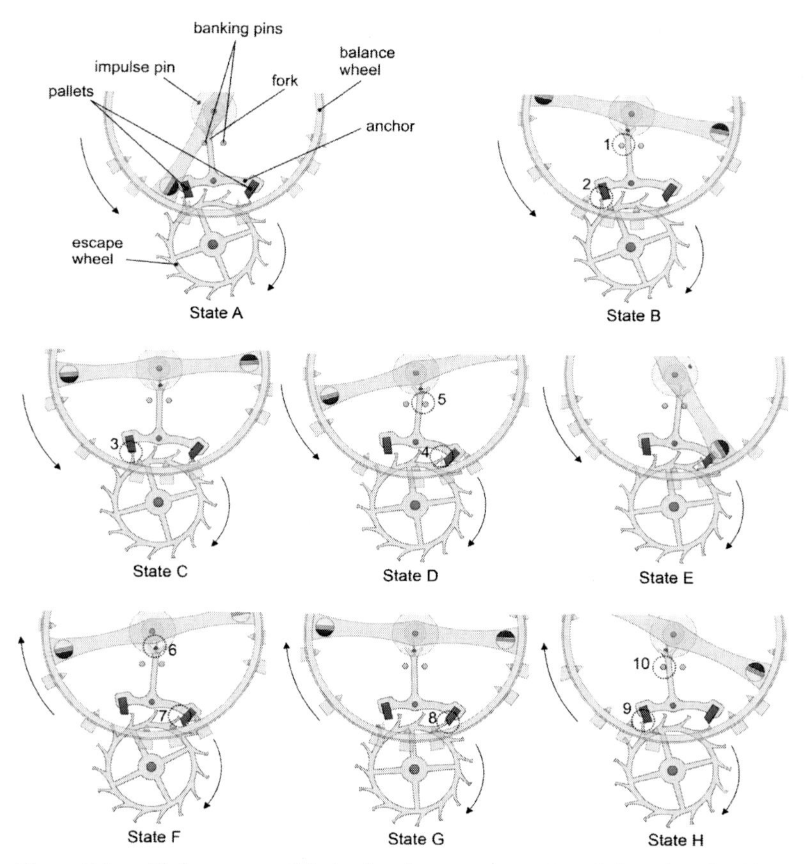

Figure 7.2 Eight states of Swiss level escapement in a full cycle. The blue doted circles annotate the shock and beat noise and sliding events [14].

After this energy impulse to the balance wheel, the escapement wheel is free again and accelerates clockwise until the next tooth hits the locking face of the right pallet (exit pallet), which in turn locks the movement (see stage D). This shock event (4) is also very critical, since with time deep wear grooves will be developed on the locking face of the pallet. A further impact (5) is between the right banking pin on the balance wheel and the fork. The lever is now held in place by the right banking pin, while the impulse pin leaves the fork (stage E). The balance wheel continues to turn counterclockwise until its spring energy is exhausted. Subsequently, the balance wheel returns in the clockwise direction, and a half cycle is completed. The same states will now be performed backwards (stages F–H). The impulse pin hits the fork (6), releasing the locked pallet (7), and the escapement wheel transfers the new impulse to the pallet (8). Finally, the tooth of the escapement wheel again locks the anchor (9), and the lever hits the left banking pin (10).

The frequency of the cycle is mainly determined by the spring components. Since the balance wheel rotates back and forth on an axis, its motion is governed by the coiling and uncoiling of the hairspring. Typically, a half cycle is called one beat, and the frequency is described by the number of beats per hour (bph). Most watches nowadays operate at 18,000, 21,600, 28,800, or even 36,000 bph.

The accuracy and lifetime of a watch are mainly determined by the constancy of single beats. The most critical components are the escapement wheel and the palettes, since about 65% of the total energy is lost due to friction and wear during the shock and sliding events. Any change in friction or minimal geometric changes due to wear grooves will inevitably lead to a frequency change and, thus, to an imprecision of time readout. Typically, these effects can be compensated with regular service intervals (every two to five years). However, a more favorable approach is the replacement of conventional materials with new and advanced technology such as NCD and, consequently, reducing friction and enhancing wear resistance.

The first implementation of CVD diamond in a mechanical watch movement was presented in 2005 with the Ulysse Nardin Freak Diamond Heart in a 99-piece limited edition (see Fig. 7.3) [15].

The escapement wheels are completely made of diamond, etched out of a microcrystalline CVD diamond wafer (see Fig. 7.1c). This production technique allows the designing of complex microparts, which cannot be machined with conventional methods. However, given the costs of the time and energy to produce these diamonds plates, it is not surprising that the parts are very expensive, and hence implementation is limited to very high-priced editions of mechanical watches.

Figure 7.3 Ulysse Nardin's Freak Diamond Heart using synthetic diamond wheels in the watch movement (Courtesy of André Flöter and Peter Gluche) [15].

An approach for a highly efficient and economical production of diamond watch parts has been developed in parallel using a process in which inexpensive silicon parts are coated three-dimensionally with diamond. This coating is so hard that it adds a high degree of stability even to the thinnest parts and still maintains the great tribological properties of diamond.

7.3 Micro-/Nanostructuring of CVD Diamond

7.3.1 All-Diamond Parts

Diamond can be patterned utilizing similar methods known from silicon technology. However, the gas chemistry and masks have to be adapted to the aggressive environment. Typically, diamond is etched in oxygen/argon plasma using a radio-frequency (RF) parallel plate or inductive couple plasma (ICP) reactors. The argon is used to bombard the diamond bonds physically, and oxygen produces the volatile compounds CO and CO_2. In contrast to deep reactive-ion etching (DRIE) of silicon, common photoresist masks cannot withstand the oxygen ambient for a long time. Consequently, other mask materials have to be used, for example, metal masks, which are also structured by photolithography. Advantage can be taken from the high anisotropy of the sputter-assisted process, and less effort has to be spent on the sidewall protection. Optimized processes yield a depth profile of more than 300 μm with a verticality of $89° < \alpha < 91°$ and a sidewall roughness $R_{rms} < 30$ nm (Fig. 7.4).

Figure 7.4 SEM picture of a structured CVD diamond micropart, showing a verticality of $89° < \alpha < 91°$ and a very low sidewall roughness. *Abbreviation*: SEM, scanning electron microscopy.

With this technique, it is possible to manufacture complex geometries with the highest precision. The initial raw material consists of large-grained (up to 30 μm), transparent CVD diamond wafers having a diameter of approximately 80 mm (3 inches) and a thickness of up to 300 μm.

Examples of microparts that are typically used in mechanical watch movements are displayed in Fig. 7.5. Both the wheel and the anchor are entirely made of optically transparent CVD diamond and enable lubrication-free escapement.

This diamond-structuring technique can also be employed using very thick, free-standing NCD. Such NCD films up to several hundreds of micrometers in thickness can be economically grown nowadays on 6-inch diameter Si wafers [16]. Figure 7.6 shows a free-standing piece of a 90 μm thick diamond plate and the corresponding SEM cross section.

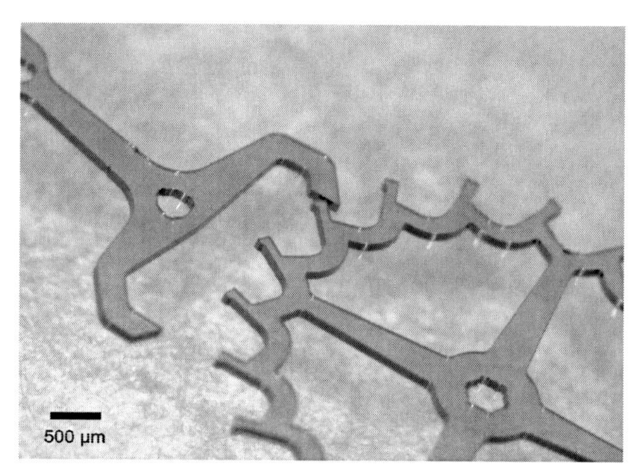

Figure 7.5 Escapement wheel and anchor made from optically transparent CVD diamond.

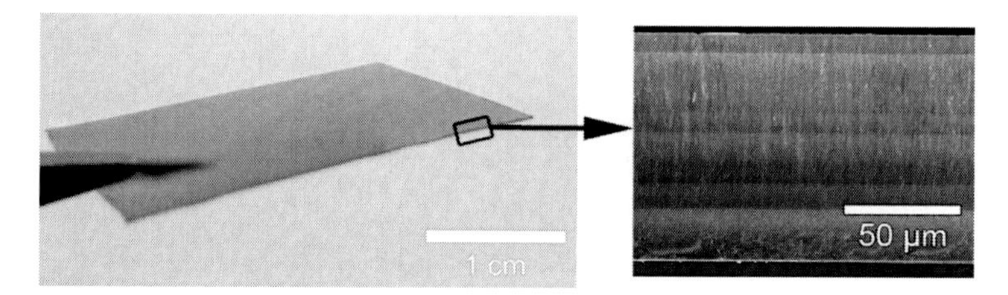

Figure 7.6 Free-standing ultrathick NCD film (90 µm) and the corresponding cross-sectional SEM picture [17].

Contrary to microcrystalline diamond (MCD) films, where a strong thickness-dependent microstructure exists and grains tend to grow columnar-like, no visible change in the nanostructure can be observed even with tens of micrometers in film thickness [17]. In consequence of the ultrasmall equiaxed crystallite structure, the roughness changes only slightly with film thickness. Ultrathick films up to 100 µm still show smooth surface roughness (R_{rms}) of less than 30 nm. Additionally, stress gradients along the growth direction are small and lead to a minimal bending of free-standing, thick coatings.

These highly isotropic NCD plates have advantageous mechanical properties (Young's modulus ≈ 700 GPa; fracture strength above 5 GPa) and are, therefore, very promising for many applications where wear resistance, high yield strength, smooth surfaces, and high load-carrying capacities are in demand.

The plasma-etching process for structuring does not differ significantly in the case of NCD. However, the resulting sidewall surface roughness is considerably smoother than the one of microcrystalline substrates. The etching of MCD typically accentuates the columnar structure of the material, resulting in vertical etching stripes, and increases the sidewall roughness, whereas in the case of NCD a minimal roughness of 20 nm can be obtained. In Fig. 7.7, a hairspring is presented, etched from a 60 μm thick NCD film.

Figure 7.7 Hairspring etched out of a 60 μm thick NCD plate having a sidewall thickness of 20 μm.

7.3.2 Diamond-Coated Silicon Parts

The above-explained technique yields to parts entirely made out of diamond. Both the diamond substrates' production irrespective of their microstructure and the subsequent etching to the desired shape still remain time-consuming and expensive steps. Accordingly, a highly economical and efficient alternative has been developed and was presented by GFD in 2006. This approach is based on a coating strategy that utilizes a cheaper substrate such as prestructured silicon and a thin, three-dimensional (3D) coating of NCD.

Silicon is the most studied and most suited substrate for diamond growth; the melting point (1,400°C) is higher than the deposition temperature window (600–800°C), and the difference of the thermal expansion is low. A conventional DRIE process is used to prestructure the device layer of commercial silicon-on-insulator (SOI) wafers. The thickness of the parts is defined by the thickness of the wafers. Prior to NCD deposition, the silicon wafers are prenucleated using a nanodiamond-seeding solution. The wafers are coated three-dimensionally, typically with 5 μm thick NCD, using a hot-filament CVD system. This thickness is chosen to make sure that high mechanical impacts like shocks are also absorbed by the coating without damage.

Figure 7.8 presents such a coated 6-inch wafer having approximately 500 wheels. The parts are each fixed by small attachments with integrated, predetermined breaking points and hence can be easily singularized after the diamond coating is completed. The cross-sectional SEM picture shows the 3D coating that perfectly follows the contour of the micropart and automatically creates radii on the sharp corners. Since the coating process is isotropic, the residual radius of curvature is equivalent to the coating thickness of the NCD film. The sidewall roughness is of high importance for the functionality of the parts, since these surfaces represent the active contact faces in the watch movement. The presented process yields reproducibly sidewall surface roughness $R_{rms} < 10$ nm (10×10 μm^2 AFM scan).

Figure 7.8 Diamond-coated prestructured Si wafer with 450 wheels (Reprinted with permission of Peter Gluche, copyright Diamaze Microtechnology SA, Switzerland) [16].

Diamond silicon parts offer many great advantages over conventionally used material escapement microparts. Typically, both the escapement wheel and the anchor are made of hardened steel, and thus, the choice of design is limited to the fabrication technique. Using the extremely well-developed DRIE process for silicon, almost any imaginable shape of such escapement components can be manufactured. This enables the integration of several parts monolithically and reduces the costs of assembly and mechanical fabrication. Figure 7.9 shows the comparison of a traditional and a diamond-coated silicon (DCS) anchor. The conventional anchor consists of four parts: the main body made of steel, two ruby pallets, and a dart made of brass. The pallets are in contact with the escapement wheel, and the active surfaces are highly stressed by shock and friction. The dart prevents the escapement wheel to enter into a forbidden condition. In the conventional anchor system, the position of the ruby pallets has to be adjusted relative to the teeth of the escapement wheel. The brass dart also has to be adjusted and finally pressed into the anchor fork.

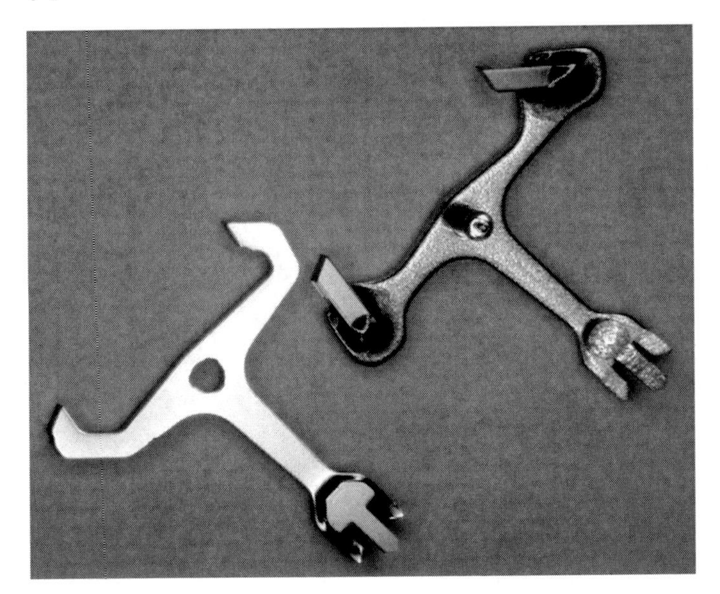

Figure 7.9 Comparison of a conventional anchor system (steel part + ruby pallets and brass dart) and a DCS anchor system (all parts are monolithically integrated). In the DCS part, the dart is directly integrated using a two-level etching technique.

Using DCS technology these parts can now be monolithically integrated, and double-side etching techniques even allow the fabrication of two-level silicon parts (see Fig. 7.9). Since the etching techniques of silicon are dimensionally very precise and enable very smooth and vertical sidewalls, adjustment of the penetration of the pallets can be avoided. To make sure that the penetration of the pallets relative to the escapement wheel is optimal (e.g., originated by tolerances of the bearing jewels of the base plate), several classes of anchors with a fixed penetration depth of pallets can be fabricated. A premeasurement of the bearing jewel distance of anchor and wheel determines which type of anchor is suited best to the movement.

The 3D diamond film around the rather brittle silicon body in turn enforces the whole structure significantly. The part may break if the applied force exceeds the diamond fracture strength, but chipping damage (e.g., by manipulation contact) as commonly present on silicon parts is never observed. In contrast to uncoated silicon parts this enables an automated assembly and simplifies the after-sales procedures such as service and cleaning. Furthermore, the shock resistance is increased considerably. The parts can even withstand shocks of 5,000 g if the general geometry is chosen correctly.

Due to the hardness of the diamond film, the contact surfaces show negligible wear. The friction coefficient of NCD versus NCD in a dry environment is around 0.07–0.01 [18], depending on their surface condition. In comparison, conventional movement (ruby/steel lubricated) shows a friction coefficient of approximately 0.12, determined by the used oil.

The combination of the light, hard, and antimagnetic silicon body with the extraordinary surface properties of NCD allows the motion in an unlubricated state and, as a consequence, significantly enhances the accuracy of the watch and reduces service intervals.

7.4 Summary

Diamond is without doubt the perfect material for any application where reliability and wear resistance are in high demand. However, diamond properties as well as crucial processing techniques have to be precisely adjusted for a given application. Today's ability to control the grain size and surface morphology of CVD diamond, combined with a highly economical 3D coating technique, offers a great new field of application.

The presented example of DCS watch parts enables the fabrication of a complex structure as well as merging of different individual parts into single components, leading to a new era of watch movement. The combination of silicon and diamond technology can be transferred to many other different applications where tribology properties are crucial, such as microtransmission systems, cutting tools, or microelectromechanical systems (MEMS). Furthermore, the isotropic growth of very thick NCD films allows the fabrication of precise structures entirely made of diamond, showing excellent surface and mechanical properties.

References

1. U.S. Geological Survey. (2011). *USGS Mineral Commodity Summaries* (Reston, USA), ISBN: 978-1-4113-3083-2.
2. The Economist Newspaper. (2007). Diamonds: changing facets (London, U.K.).
3. CemeCon. (2008). Milling of filigree contours with μm-accurately tolerances, *CemeCon Facts*, **31**, pp. 6–7.
4. Weigand, M. (2011). Machining CFRP with diamond, *CemeCon Facts*, **36**, pp. 6–7.
5. Flöter, A., Gluche, P., Bruehne, K., and Fecht, H.-J. (2007). Diamond coat hones the cutting edge, *Metal Powder Rep.*, **62**, pp. 16–20.
6. "CVD diamond heatspreaders." Available at http://www.cvd-diamond.com/applications_en.htm. (Accessed November 11, 2011).
7. "Optical windows." Available at http://diamond-materials.com. (Accessed November 11, 2011).
8. Sumant, A.V., Grierson, D.S., Gerbi, J.E., Birrell, J., Lanke, U.D., Auciello, O., Carlisle, J.A., and Carpick, R.W. (2005). Toward the ultimate tribological interface: surface chemistry and nanotribology of ultrananocrystalline diamond, *Adv. Mater.*, **17**, pp. 1039–1045.
9. Advanced Diamond Technologies. (2010). *UNCD Faces Whitepaper–2010* (Romeoville, USA).
10. Advanced Diamond Technologies. (2011). *NaDiaProbes®—All-Diamond AFM Probes* (Romeoville, USA).
11. Fu, Y., and Du, R. (2007). "Swiss lever escapement mechanism," in *Mechatronic Systems: Devices, Design, Control, Operation and Monitoring*, ed. De Silva, C.W. (CRC Press, Boca Raton), pp. 3.1–3.17.

12. Tam, L.C., Fu, Y., and Du, R. (2007). Virtual library of mechanical watch movements, *Comput.-Aided Des. Appl.*, **4**, pp. 127–136.
13. Witschi Electronic. (2010). *Test and Measuring Technology Mechanical Watches* (Büren a.A., Switzerland).
14. "Die Schweizer Ankerhemmung." Available at http://uhrentechnik.vyskocil.de/56.0.html. (Accessed November 9, 2011).
15. Flöter, A., and Gluche, P. (2005) Erstes CVD Diamantzahnrad in der Hemmung eines Uhrwerkes, *Industriediamantenrundschau IDR*, **39**, pp. 110–112.
16. Diamaze Microtechnology SA, Rue des Crêtets 138, 2300 La Chaux-de-Fonds, Switzerland.
17. Wiora, M. (2013). *Characterization of Nanocrystalline Diamond Coatings for Micro-Mechanical Applications*, Doctoral dissertation, Ulm University, Germany.
18. Wiora, M., Sadrifar, N., Bruehne, K., Gluche, P., and Fecht, H.-J. (2011). Correlation of microstructure and tribological properties of dry sliding nanocrystalline diamond coatings, *Tech. Proc. NSTI Nanotechnol. Conf. Expo. NSTI-Nanotech 2011*, **2**, pp. 293–298.

Chapter 8

Economic Analysis of Market Opportunities for CNTs and Nanodiamond

Matthias Werner,[a] Mario Markanovic,[a] Catharina-Sophie Ciesla,[a] and Leif Brand[b]

[a]*Nano and Micro Technology Consulting (NMTC), Soorstr. 86, 14050 Berlin, Germany*
[b]*VDI Technologie Zentrum GmbH – Zukünftige Technologie Consulting, Air Port City, VDI Platz 1, 40468 Düsseldorf, Germany*
werner@nmtc.de

Carbon is the sixth element in the periodic table and of the same importance to nanotechnology as is silicon to electronics. Even though carbon holds the 17th place in terrestrially available elements, with a portion of 1:1300 to silicon as the second most terrestrially available element, carbon is one of the most important elements at the same time since it is essential for the structure of any organic material. Its midposition in the periodic table allows interaction with both electropositive and electronegative reactants to form stable substances. The multitude of allotrope carbon materials partly shows totally opposing characteristics, making carbon one of the most interesting materials in current research. These materials are

Carbon-based Nanomaterials and Hybrids: Synthesis, Properties, and Commercial Applications
Edited by Hans-Jörg Fecht, Kai Brühne, and Peter Gluche

ISBN 978-981-4316-85-9 (Hardcover), 978-981-4411-41-7 (eBook)
www.panstanford.com

in focus of intense research and promise numerous applications in the fields of electronics, medicine, and diverse other nano-enabled technologies. One of the hot topics in carbon-based materials is still carbon nanotubes (CNTs) and nanodiamond. Despite their major technological advantages to classical materials mass market applications for these novice materials are not existent yet. The main reason is their high price caused by low availability. However, niche markets could prove to be drivers for scaling up nanomaterials if certain quantities can be produced to beat down the price and hence enable the use of carbon-based nanomaterials in other areas. Apart from the scientific and technological aspects this chapter mainly discusses the economical chances and barriers of CNTs and nanodiamond. In this context also the technological readiness level of possible application fields is discussed, and a general market overview with corresponding drivers and barriers is given.

8.1 Introduction

Carbon-based materials and especially carbon nanotubes (CNTs) represent one of the most research intensive areas within nanotechnology. Their extraordinary material properties have given rise to numerous application scenarios since the early 1990s. These include new composite materials, electrode coatings, energy storage, sensors, alternative flat-panel display technologies, novel nanoelectronic components, pharmaceutics, and biomedical applications. Due to the wide range of applications, CNTs are relevant for many industries.

In many areas, CNTs are still in an early research stage; however, recently first applications have been leaving the development laboratories. Above all, mechanically reinforced CNT plastics are increasingly being commercialized. Often, these are lifestyle and sports products in which the reference to new high-tech materials can be effectively advertised. Also, the production of electrically conductive composites, coatings, lacquers, and paints is gaining importance, and these are flowing into marketable products.

So far CNT-based products have largely addressed niche markets; however, due to increased industrial production capacities, CNTs are decreasing in cost, stimulating further expansion, particularly in the composite sector.

In general, CNTs offer considerable innovation and commercialization potential. Some years ago Europe was in danger of falling behind international competitors. However, due to large research funding projects stimulating innovation and numerous industries expanding their production capacities, there is now the opportunity to overcome such gaps. The associated stimulation effect opens up possibilities for Europe, not only to compete effectively with the United States and Asia, but also to establish technological leadership positions in particularly crucial areas and to achieve rapid and targeted implementation into commercial products and applications [1].

8.1.1 Definition

CNTs are nanoscale cylindric tubes that consist of rolled-up layers of graphite. The tube ends may remain open but are often capped with a hemisphere of the fullerene structure. Their diameter ranges from ~1 nm to ~50 nm, and their length is typically in the micron range but can reach several millimeters. Single-walled carbon nanotubes (SWNTs or SWCNTs) and multiwalled carbon nanotubes (MWNTs or MWCNTs) are distinguished. SWCNTs can be formally derived from a seamlessly rolled-up graphite layer. MWCNTs consist of multiple concentrically aligned single-walled tubes of different diameters. A specific configuration frequently appearing in the literature is MWCNTs consisting of exactly two concentric nanotubes. They are called double-walled carbon nanotubes (DWNTs or DWCNTs).

8.1.2 Overview

8.1.2.1 Carbon nanotubes

Ideal CNTs show a number of extraordinary mechanical, chemical, thermal, electrical, and physical properties, making them one of the most promising materials within nanotechnology.

CNTs are characterized, in particular, by the following exceptional properties [2, 3]:

(i) **Electrical conductivity**: There is a "ballistic" electron transport in metallic CNTs along the tube axis; upon moving through the transport medium electrons are not scattered on length scales of several microns. Due to quantum mechanical effects there are nearly no collisions with the carbon atoms in the lattice structure. The achievable electric current densities are more than 1,000 times higher than those of metal wires made of copper or silver.

(ii) **Thermal conductivity**: With >3000 W/(mK) the thermal conductivity of MWCNTs considerably exceeds that of diamond, 2,300 W/(mK), or silver, 430 W/(mK).

(iii) **Mechanical properties**: The elastic modulus of MWCNTs is given in the literature as ranging from 0.2 TPa to 1 TPa and the tensile strength between 10 GPa and 60 GPa. Thus the tensile strength normalized to the weight is between 60 and 360 times greater than that of steel.

(iv) **Aspect ratio**: CNTs have a very high length-to-diameter ratio, making them prone to buckling under compression. However, this high aspect ratio is ideal for electron emission applications such as cold field emission in displays or X-ray tubes.

(v) **Chemical reactivity**: Due to their large surface area and bending of the surface, CNTs are chemically quite reactive compared to graphite monolayers (graphene). This allows for a chemical modification of the CNT surface, which can affect, for example, the solubility of CNTs in different solvents. In addition, the CNT surface may be functionalized, which is important for applications such as the biomedical sector, electronics and sensors, and composites processing.

Concerning their electronic properties, SWCNTs appear in fractions of conducting (i.e., metallic) and semiconducting nanotubes. The conductivity depends on the structural setup (i.e., the chirality or "roll-up direction") of the tubes. Metallic tubes allow for extraordinarily large current densities due to their high carrier mobilities and at the same time show only low electrodegradation of the material. Semiconducting tubes, on the other hand, give rise to the fabrication of nanoelectronic transistors, which have already been realized as prototypes. Due to their electronic properties and the appearance of both metallic and semiconducting tubes, CNTs are discussed as a promising material in the postsilicon era in microelectronics and data storage.

The electronic properties of CNTs also give rise to a number of other applications such as electron emitters in new flat-panel display types, electrode materials, fillers for antistatic and conductive plastic composites, electrode materials in electrochemical applications, etc.

On the basis of their chemical properties exposing a large and reactive molecular surface, numerous biochemical functionalization of CNTs have been realized, making them candidates for compound tracers or drug delivery systems in the biomedical area. However, research is at a very fundamental stage in this sector, and issues of biocompatibility and toxicity remain open at the moment.

Significant effort has been spent in the development of large-scale production techniques. Meanwhile a series of industrial production plants have been established, leading to a decline in CNT prices [4] and giving rise to more widespread applications. The main consumption will be due to utilization in novel nanocomposite materials. A number of products based on such composites have already been commercialized, such as automotive fuel pipes and car body parts but predominantly for lifestyle products, including tennis and hockey rackets, bicycle frames, and skiing equipment.

CNTs in their various forms are currently at the market entry level at least in terms of individual applications. The transfer from lab-scale fabrication to mass production of the raw material has very recently been realized, and an increasing number of commercial products contain CNT materials. In addition to composite materials this is also true for electrode materials. However, many of the CNT applications discussed here have not yet been released to market; they are in different states of research, ranging from fundamental research to prototypes and lab models.

8.1.2.2 Nanodiamond

Metal–diamond galvanic coatings, polishing pastes, lubricating oils and coolants, polymer composites, and greases belong to traditional ultrananocrystalline diamond (UNCD) particulate applications.

Using common galvanic equipment electrochemical deposition of UNCD, together with metals, has been proven to be useful for coatings of components of transportation units, tools for electronics, watches, medicine, and the jewelry industry. The addition of UNCD to galvanic coatings leads to the following advantages:

- increased wear resistance (2 to 13, times depending on the metal);

- microhardness (up to 2 times, depending on the metal);
- increase in corrosion resistance (2 to 6 times, depending on the metal);
- decrease in porosity (pores can be completely eliminated depending on the metal);
- significant decrease of the friction coefficient;
- considerable improvement of adhesion and cohesion; and
- high throwing power of the electrolyte.

According to interviews [5] the service life of products is increased 2 to 10 times, even if the coating thickness is decreased to a half or a third. In many metals that are employed in electronic applications, such as silver, gold, and platinum, the effect of strengthening can be noticed. Mostly in strengthening chromium coatings an electrolytic process UNCD is used, in particular. Thereby there is no need for any modification of the standard production line when adding UNCD-containing additives to the chromeplating electrolyte. The operating life of molds, high-precision bearing surfaces, and other similar components is higher when using such coatings. The UNCD percentage in a metal coating ranges from 0.3 wt% to 0.5 wt%; 0.2 g/m^2 (1 $carat/m^2$) of UNCD is required for a 1 mm thick metal layer. Finishing precision polished materials for electronics, radio engineering, optics, medical, and jewelry industries is the field of application for such UNCD particles. Composites including UNCD allow for complex geometrical surfaces with a relief height roughness of 2 nm to 8 nm.

8.1.2.3 Comparison of carbon nanotubes and nanodiamond (Table 8.1)

CNTs have a diameter of a few nanometers and due to the unique structure have many exciting properties. However, their physical properties are partly comparable to those of diamond.

It is interesting to note that diamond, in terms of physical properties, shows some similarities to CNTs. The Young's modulus of diamond in the <100> direction of 1,050.3 GPa is close to that of CNTs. This is similar to the maximum tensile strength of nanotubes, which is reported to be 45 ± 7 Pa compared to at least 60 GPa for diamond. Furthermore, the experimentally reported thermal conductivity of MWCNTs ranges widely, from 25 W/(mK) to more than 3,000 W/(mK) and corresponds to that of high-quality CVD diamond or the best natural-type IIA diamond. It should be kept in mind that in the case of diamond nanoparticles, boundary scattering will limit the thermal conductivity for grains in the nanometre range. Theoretical predictions of the thermal conductivity of CNTs indicate that the thermal conductivity can be as high as 6,600 $Wm^{-1}K^{-1}$ at room temperature. However, if CNTs are used as a filler material for the improvement of the thermal conductivity of a composite, one can expect significantly lower thermal conductivity.

Table 8.1 Some physical properties of single-crystalline diamond, CNTs, and nanodiamond

Physical properties	Single-crystalline diamond	CNTs	Nanodiamond
Grain size (nm)	up to 10^7	–	5–100
Thermal conductivity ($Wcm^{-1}K^{-1}$)	22	60	0.06–0.1
Minimum resistivity p-type (Ωcm)	$<10^{-3}$	Theoretically length independent (but high contact resistance at interfaces)	$<10^{-3}$
Fracture strength (GPa)[a]	4	~100 (theoretical value)	<5
Vickers hardness (GPa)	120	–	88
Young's modulus (GPa)[b]	1,143	>1,000	~1,000

[a]The theoretical value of UNCD fracture stress (σ_f = 100 GPa) is very large compared to that observed experimentally (σ_f < 5 GPa). The fracture stress is strongly dependent on defects. Even natural single-crystalline diamond is notoriously inhomogeneous.

[b]The Young's modulus of single-crystalline diamond is slightly dependent on the crystal orientation.

8.2 Scientific and Technological Aspects

To enable a comparable evaluation of nanotechnology-related applications, the following "technology readiness levels" (TRLs) have been defined for scientific or technological evaluations within "ObservatoryNANO" reports:

- TRL 1: "basic research"
- TRL 2: "applied research"
- TRL 3: "prototype/lab device"
- TRL 4: "market entry"
- TRL 5: "mature market" (mass or established niche markets)

TRLs will be used to classify specific applications according to their technological readiness.

8.2.1 State of R&D

Due to their outstanding mechanical, electrical, thermal, and chemical properties CNTs are considered as promising material in nanotechnology, with a wide range of potential applications. However, despite intensive research only few CNT-based products have reached the stadium of commercialization, and the majority of potential applications are currently still in the research or prototype state. The situation for different areas of application varies significantly.

8.2.1.1 Nanoelectronics (TRL 1)

On the basis of their molecular size and electrical properties nanotubes are ideal building blocks for nanoelectronics. In a variety of experiments, CNTs have been demonstrated as active electronic devices such as diodes and transistors, as passive interconnects, or as data storage units. All electronic applications of this kind are, however, still at a very early stage of research. While the functionality of individual laboratory models has been widely shown, a transfer into industrial mass production is currently not foreseeable. One of the main difficulties is the fabrication of pure and homogeneous nanotubes with defined conducting properties and their mass production–compatible targeted alignment and contacting on substrates. The further development of lithographic methods and the development of targeted growth processes of SWCNTs on Si substrates are crucial for advances in CNT nanoelectronics. A complete replacement of silicon-based electronics by CNT technologies is currently not foreseeable. This is due to the mentioned difficulties of CNT electronics and to the profound degree of maturity and variability of the established silicon technology.

8.2.1.2 Biomedical applications (TRL 1 to TRL 2)

Biomedical applications of CNTs have gained considerable interest. CNTs have, for example, been demonstrated as molecular transporters for genes, proteins, and pharmacological drugs ("drug delivery"). Further applications as implant materials or in purification membranes are also possible. However, the majority of possible applications are currently still at the fundamental research stage. Various biomolecular and technological difficulties have to be overcome prior to a mature medical or therapeutic utilization. However, due to Europe's important position in medical research and medical engineering biomedical applications of CNTs have promising commercial potential for the future.

8.2.1.3 Biocompatibility (TRL 2)

The issue of biocompatibility or toxicity is of crucial importance for the use of CNT-based materials, in particular in the biomedical area. The first systematic studies on the biocompatibility of carbon nanomaterials have been conducted in recent years. However, the results are not yet definite, and concluding statements are still not possible. In principle a lung hazard effect of free CNTs can be assumed. Their hazard potential may be evaluated as lying between other carbon materials, such as carbon dust, soot, or fullerenes, and the more dangerous asbestos. Cytotoxic effects of CNTs have been shown in cell biological studies; however, various experiments show that the risk is strongly dependent on parameters such as the shape, length, and stiffness of CNTs used for a specific application. Beyond this the question of whether dangerous effects are due to the nanotubes themselves, to impurities contained in them, or to fiberlike agglomerations remains open. For end users of CNT products hazards may presently be limited. For people working in the immediate environment of CNT production, the hazard potential is similar to that of manufacturing plants of paints and solvents or to the exposure to dusts and fumes; the necessary precautions are, therefore, similar to the handling of other potentially hazardous substances.

8.2.1.4 Hydrogen storage (TRL 2)

Initial expectations regarding the hydrogen storage potential of CNTs have not been realized. The economic relevance threshold of 6.5 wt% of stored hydrogen, as defined by the US Department of Energy, has not yet been achieved despite intensive research. Overall, activities in the area of the CNT-based hydrogen storage have declined significantly and the use of CNTs as a medium-term commercial hydrogen storage appears unlikely.

8.2.1.5 Displays (TRL 3)

CNT-based field emission displays (FEDs) are currently at an advanced prototype stage. Market launches have been announced several times; however, due to technical implementation difficulties, especially in mass production, they have repeatedly been postponed. CNT displays are addressing the rapidly growing market for flat screens, which is currently fully dominated by liquid crystal displays (LCDs) and plasma displays. CNT-based displays are seen as promising but are in commercial competition with the more established LCD technology. The latter has made great technological progress, especially during the last few years, and now occupies a dominant market position. In addition, organic light-emitting diodes (OLEDs) are a promising future development; with respect to CNT displays OLEDs may be regarded as a competing technology. CNT-based displays have only minor commercial relevance for Europe, as major industrial producers, and associated research and development (R&D) departments and production lines, are largely located in the United States and especially in East Asia.

8.2.1.6 Sensors (TRL 3)

Sensor applications of CNTs are evaluated as very promising. CNT-based heat and gas sensors as well as chemical and biological sensors have been successfully demonstrated; however, so far only lab devices have been developed and broader commercialization is still lacking. The latter requires uniform CNTs as well as their exact positioning and electric contacting on the substrates. Hence, current drawbacks in the area of CNT-based sensors are similar to those in nanoelectronics.

8.2.1.7 Electrochemical applications (TRL 3 to TRL 4)

Commercial uses have also been identified for electrochemical applications. Here, CNTs are increasingly utilized as electrode materials in batteries, supercapacitors, and fuel cells. New electrode materials play a crucial role due to the growing importance of emerging battery and fuel cell technologies, the increasing demand for efficient energy storage for portable devices, and also the increasing demand for automotive and propulsion technologies.

8.2.1.8 Composite materials (TRL 4)

Commercialization is most advanced in the area of polymer composites. Here the most obvious commercial uses with the highest immediate market potential, especially for Europe, can be seen. The first products, for example, from the sports and lifestyle sector, have already been on the market for some time; however, these are still too expensive to compete successfully in the mass market against materials such as carbon fiber–reinforced composites. With the expected continuing decline in the price of CNT raw material and CNT-based composites a considerable potential will be opening up in this area. Besides mechanically reinforced composites a great potential can be seen, particularly for electrically and thermally conductive polymer composites.

8.2.1.9 Production of polymer composites (TRL 4)

A major challenge concerning the economic exploitation of the positive properties of MWCNTs is their occurrence in the form of agglomerates, which often prevents good dispersion in composites. Recent research expects that neither CNT agglomerates nor short, mechanically sheared CNTs may transform the desired positive properties of nanotubes into the composites. Hence, it is also the processing of the composites that plays an important role. The procedure consists of two steps: first, on the basis of chemical surface functionalization or by use of suitable dispersants, the CNTs have to be prepared in a way that is principally allowing for dispersion within the compound. For the utilization of nanocomposites in an application, the second step requires the adjustment of the recipe to the specific property profiles through the use of further additives, fillers, processing aids, stabilizers, etc.

Due to the numerous interactions between the polymer matrix, functionalized CNTs, and further additives and auxiliaries, complex processes and lengthy formulation tasks are mostly required.

In principle there are three ways for the production of thermoplastic matrix composites:

- mixing of CNTs in polymer solutions and pouring as (thin) films;
- in situ polymerization of polymer monomers in presence of CNTs; and
- compounding of thermoplastics with CNTs via master batches or direct mixing processes.

For industrial applications compounding with existing techniques is of particular importance. For mixing mainly twin-screw extruders are used, but even alternative extruding techniques are utilized [6].

The first practical implementations of thermoplastic polymers have been successfully carried out by Leistritz Extrusion Technology (Germany). They used corotating twin-screw extruders, since the required shear can be well realized. As a result of shear stress, the CNT primary agglomerates are broken and the majority of the CNTs are dispersed in the polymer. It turned out that the quality of the dispersion of CNTs within thermoplastic polymers depends on several parameters:

- screw stocking (kneading elements);
- extrusion speed;
- CNT concentration (>10% favorable); and
- polymer material (e.g., CNTs more easily dispersed in polycarbonate than in polyactide acid).

Components made of CNT composites can be produced either directly from the extruder or by injection molding. The incorporation of CNTs into different polymers may be made either by a compound or by a master batch [7].

8.2.1.10 Technical production of carbon nanotubes (TRL 5)

Meanwhile, CNTs can now be produced on an industrial scale; there are a number of sources of CNTs available at a reasonable cost per unit [8]. MWCNTs are predominantly produced as for these a high purity level is achievable on an industrial scale. Due to increased production capacities and profitable production processes the price for MWCNTs is in the range of 0.10–0.15 €/g, depending on material quality and acquired quantities. For highly pure SWCNTs, in contrast, prices still are in the range of 100 €/g [9, 10]. Over recent years the price decline for MWCNTs has exceeded a factor of 100. In the case of further enhanced production capacities, prices of below 0.04 €/g for MWCNTs are likely on a medium run [11]. SWCNTs are more expensive to produce than MWCNTs. Meanwhile the complex manufacturing process of SWCNTs has been transferred into a technical scale (e.g., by Nanocyl or Thomas Swan). The price of SWCNTs is expected to fall by a factor of 10 with increased production volumes.

Major CNT producers such as Bayer MaterialScience (Germany), Arkema (France), Hyperion (U.S.), Nanocyl (Belgium), Iljin Nanotech (South Korea), Shenzhen Nanotech Port (China), and Japanese companies [12] predominantly use catalytic CVD for large-scale MWCNT production. For SWCNT production also laser ablation, arc discharge, or high-pressure CO processes are utilized.

In 2009, Bayer MaterialScience established a production plant with a capacity of about 200 tons per year. A production capacity of 3,000 tons per year is the target for the medium term. MWCNTs are processed as agglomerates, which are still contaminated with catalyst residues, and have to be "unravelled" before further processing; however, as they are not dusty, they permit easy handling. Bayer MaterialScience is currently expanding its sales and cooperation network. Besides its distributor, Brenntag Schweizerhall, the company has recently completed an agreement with Toyota Tsusho on exclusive distribution rights for its product "Baytubes" in Asia.

Very recently Arkema (France), which started its CNT activities in 2003, announced the construction of a CNT pilot production plant with a capacity of 400 tons per year starting from 2011 [13]. As a novelty, this plant will be the only CNT production site in the world that will use an entirely biosourced raw material. Arkema started its first pilot laboratory capable of producing some 20 tons per year of CNTs in 2006. Meanwhile it developed a range of innovative master batches that are easy to process within various thermoplastic, elastomer, and, more recently, thermoset matrices. These master batches help to optimize the application properties of end products.

Bayreuth-based Future Carbon (Germany) goes one step ahead in the value chain. The company also offers products in the form of CNT dispersions and functionalized CNTs for further processing in respective applications [14]. Even other companies and institutes, such as Amroy, Nanocyl, Hyperion, EMPA, and Polymaterials AG, are increasingly investigating industrial-scale processing of functionalized CNTs in order to provide reasonable prices for industrial end producers [15].

8.2.2 Additional Demand for Research

Within the nanomaterial family, CNTs probably represent the most prominent and the most promising species. In some respect they are even representative of the whole range of carbon-based nano-materials. However, there is a different R&D state ranging from fundamental research to market maturity for CNTs, depending on the application. A multitude of open questions and a variety of research items remain to be addressed.

In general, CNT applications, even within the industrial sector, are expanding rapidly with the first mass production plants now installed. The predominant and most widespread applications are in the composite material sector; however, there are a large number of other applications for CNTs at varying stages of development. The most critical applications with respect to the utilization of CNTs are the ones demanding a high purity of specific nanotube types. For electronic applications, for example, it is crucial to have either pure semiconducting or pure metallic lots of SWCNTs; however, manufacturing pure CNTs remains a challenge. This is also true for the controlled arrangement or controlled deposition growth of CNTs.

Whereas CNT-based nanoelectronics is still at an early stage, hydrogen storage is somewhat beyond the hype. Original expectations to store significant amounts of hydrogen in CNT materials have not been realized.

Another promising application is the flat-panel display sector. CNTs are discussed as one of the future technologies. First prototypes have already been presented; however, efficient display sealing in mass production has not been achieved yet, and even homogeneity of large-area CNT distribution is still an issue. In addition, among future display technologies, CNT-based flat panels face strong competition from both the ever-improving LCD technology and the promising OLED approaches. Research effort and funding spent in CNT display development has been drastically reduced recently. However, it is mainly Asia that is concerned by this decline, as there are no major display industries located in Europe.

Further demand for intense research may also be seen in biomedical applications. Numerous applications discussed are temporally far ranging, and a number of challenges remain to be addressed. One of the most crucial items is biocompatibility. After all biocompatibility and toxicity of nanomaterials, and in particular CNTs, remain an issue and are up to further investigation.

The commercial realizations and economic potential of CNT-based applications appear different depending on the particular application. In general, there is considerable additional R&D demand for all areas. Major hurdles for numerous applications are currently lacking purity and homogeneity as well as the difficult production reproducibility of the fabricated raw material. The further development and optimization of suitable analysis methods for quality control is urgently required. In addition suitable mass production compatible alignment and configuration and contacting methods are still lacking, which is a particular drawback for nanoelectronics and sensor applications.

Another major challenge is the suitable dispersion of CNTs, in particular in composite materials and in display technology.

8.2.3 Applications and Perspectives

Applications for CNTs are numerous. According to their stage of development they are already utilized as or discussed for:

- matrix materials or fillers in composites (large-scale utilization, mechanical reinforcement, conductive and antistatic materials/polymers) (TRL 4 to TRL 5);
- lightweight construction (TRL 4);
- electronic devices (large-area flat-panel displays, nanoelectronic transistors, logic circuits, computer memory, quantum information technologies, etc.) (TRL 1 to TRL 4);
- optoelectronics (photonic crystals, solar cells) (TRL 1);
- energy storage (hydrogen storage, electrode materials for supercaps, batteries, fuel cells, etc.) (TRL 3 to TRL 4);

- sensors (gas, heat, chemicals, etc.) (TRL 3);
- actuators (e.g., artificial muscles) (TRL 3);
- medical applications (implants, probes, prostheses, orthopedic equipment, surgical tools, "drug delivery systems," etc.) (TRL 1 to TRL 2);
- nanoscience nanotechnology (atomic force microscope [AFM] tips, molecular tweezers, etc.) (TRL 2); and
- chemistry (catalysts, gas adsorbents, etc.) (TRL 3).

8.2.4 Current Situation within the EU

Within the European Union (EU) numerous research institutions and groups are involved in CNT R&D. From a qualitative view they frequently belong to the front end of worldwide research in their respective fields. However, the quantity of research as well as the transition from research to commercial applications often appears much weaker when compared to other world regions. Taking the situation of CNT research and applications, the number of scientific publications and patent applications as well as of public funding activities ranges far behind those of the United States and Japan.

European companies appear to be clearly reserved concerning both the utilization of carbon-based nanomaterials and investments into appropriate R&D. Whereas in the electronics sector, including displays, Japan and other East Asian countries, as well as the United States, are fundamentally dominating applied and transfer research, the European position appears to be more promising in the areas of energy storage, environmental applications, composite materials, and nanotoxicity research. Meanwhile, a number of European companies have shown an increasing interest in new electrode materials as well as in carbon-based composite materials. In recent years the first large-scale CNT production plants have been installed and scaled up. The production capacity meanwhile has been enhanced to several hundred tons per year.

The quality of CNT research in Europe is consistently assessed as being highly valuable. However, research activities are more extensive in the United States and East Asia. These situations, in addition to the technological barriers described above, are currently addressed by larger funding projects for applied research in European states. Examples for these are the "Genesis" program in France or the "Innovation Alliance CNT" in Germany. Thus, further progress and breakthroughs as well as an improvement of the patent situation may be expected from European CNT activities in upcoming years.

8.3 Economic Aspects

8.3.1 General Market Description

This subsection deals with two types of carbon-based materials, CNTs and nanodiamond. CNTs and metal oxides are mentioned to have the best prospect of success concerning nanomaterials in nanotechnology in 2009 to 2014 [16].

CNTs have always been considered as a future mass market for several applications. However, there has not been a significant commercial breakthrough up to now. Commercially available market studies and future prognoses are most likely science driven and overoptimistic. At least in some fields, for example, microelectronics, it is very questionable if there will ever be a significant market for CNTs. Furthermore, technological problems, such as control of the chirality and therefore the required physical properties, remain to be solved. A similar discussion was ongoing 20 years ago with polycrystalline diamond films.

Today, it appears that polycrystalline diamond can serve some niche markets only, even when polycrystalline diamond films have extreme physical properties, which are at least comparable with diamond, when compared with conventional materials. One key issue of both diamond films, including nanocrystalline diamond films, and CNTs is the exorbitant price of these materials. This was almost overlooked by the scientific community, which was most likely blinded by the excellent physical properties of CNTs.

For the sake of comparison, the experimentally reported physical properties of diamond and SWCNTs in relation to silicon are shown in Fig. 8.1. According to this comparison of physical properties of diamond with SWCNTs, it does not seem appropriate to talk about better properties of CNTs. Only concerning the diameter-to-length ratio and the lower density, nanotubes are superior compared to diamond. This leads to the conclusion that the variety of applications where the use of nanotubes exceptionally promises major advancements is limited to a few applications. Examples are reinforcement of polymers or rubber, scanning probe microscope (SPM) tips, or emitter materials for FEDs.

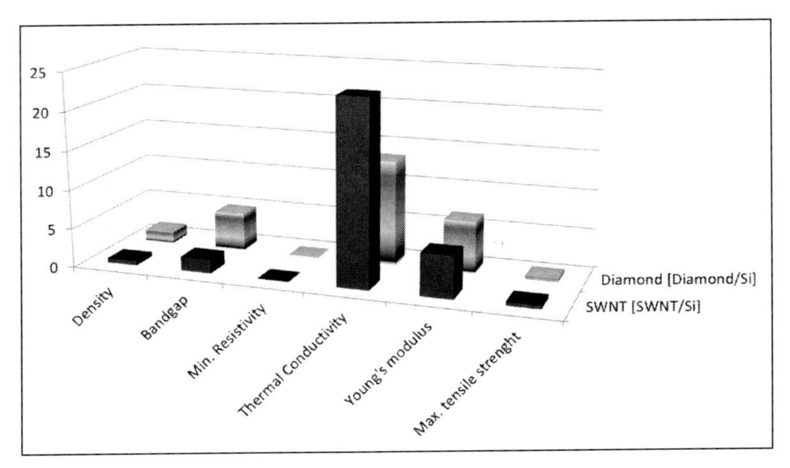

Figure 8.1 Selected physical properties of CNTs and diamond normalized to the properties of silicon (*Source:* NMTC).

Depending on the quality of CNTs in terms of purity, defects, single walls, or multiple walls, current prices of CNTs vary roughly between 1.5 USD/g and 1,000 USD/g, with an average price of 180 USD/g for SWCNTs and 0.10 €/g for MWCNTs compared to that of synthetic diamond, ranging between 0.5 USD/g and 7,500 USD/g, depending among other things on the fabrication method, quality, and size of the crystals (Table 8.2). Most of the discussed applications do not take into account the high price of CNTs or possible alternative technologies. Therefore, a large number of possible applications will not be possible due to the high price of CNTs.

Table 8.2 Material costs of SWCNTs and MWCNTs in comparison to carbon black and gold—only average prices considered (*Source:* NMTC)

	SWCNT	MWCNT	Carbon black [17]	Gold [18]
Price (USD/g)	180	0.135	0.001	28

According to a market forecast by BCC Research concerning CNTs [19] nearly 51 million USD was spent on CNTs in 2006. This value rose up to about 80 million USD in 2007. Four years later, in 2011, it was estimated that 807.3 million USD could be reached.

8.3.2 Drivers and Barriers

Looking at the value chain in the nanomaterials sector, nanotechnology has a big influence in materials development [5].

Drivers of the materials industry are clearly chemical companies. Through the controlled size-selective synthesis and assembly of nanoscale building blocks a change and manipulation of nearly all material properties can be done in a very strong way. However, the wide-ranging synthesis approaches have a number of key challenges that may be summarized as follows:

- the ability to scale up synthesis and assembly strategies for low-cost, reproducible, large-scale production of nanostructured materials, while maintaining control of critical feature size and quality;
- control of the size and composition of both nanodiamond and CNTs; and
- control of the interface and distribution of nanocomponents within the fully formed materials.

However, the most significant problem is the potential health and environmental risks of nanomaterials. Currently, very little is known about the pathways into the human body and the possible impact of nanomaterials on health. Up to now several kinds of nanomaterials like fullerenes are used in commercially available products such as cosmetics. The impact on health and the environment is not sufficiently clear yet, and this is the reason why the German Federal Environmental Agency (UBA) has most recently given a warning against using products with nanoparticles inside until the effects on human health are not better explored [20]. This overreaction has produced a lot of protest from the nanocommunity and other ministries. However, if the public is not completely informed this could have an enormous negative impact on nanotechnology-related products.

An international research team investigated how carbon-based nanoparticles interact with cells. They found strong biophysical evidence that nanoparticles may alter the cell structure and pose health risks, depending on the exposure conditions and the interaction between nanoparticles and other compounds in the human body [21]. This can also have significant implications for the commercialization of products. As long as the consequences of using nanomaterials in commercially available products are unknown some industrial players have serious reservations to use these materials [5] in products, even when nanomaterials for specific applications promise better performance. The same properties that nanomaterials are designed for may cause health and environmental problems. One example is that with decreasing particle size the surface-area-to-mass ratio becomes greater. Therefore, the specific surface and reactivity increase. This property is desired, for example, in the case of catalysts, but it can also lead to greater toxicity for living organisms.

Due to a lack of information, up to now there are many uncertainties as to under which conditions nanomaterials are likely to pose health and environmental risks. From this arises an important hurdle for the commercialization of nanotechnology-related products. In addition, other commercialization aspects can also represent significant hurdles.

Another current challenge is the economic/financial crisis, which can be assumed to be far from over. Harper summarized five possible influences of the economic crisis on nanotechnology in a white paper [22]. He has given arguments that especially venture capital funding of nanotechnology start-ups is thin on the ground. Besides a number of possible negative effects, the white paper points out positively that now many companies have a clear market focus and address real and critical needs in a cost-effective manner.

Recent survey results [23] of European micro-, nano-, and materials enterprises indicate that almost half of the responding small and medium-size high-tech companies were affected by a decrease in orders and in sales. About two-thirds of the responding companies were expecting negative effects on their businesses development in 2009.

At the current stage it is virtually impossible to predict all consequences of the economic crisis on nanotechnology-related products. This includes also the nanomaterials sector. According to the authors' view, it can be expected that future development will slow down further and will not only affect small and medium-size companies but also global players.

Potential health risks that people are concentrating on appear to be a substantial barrier to CNTs [5]. CNTs had started to be processed and synthesized in massive amounts, but being long, thin, and biopersistent, it is now assumed that they can be carcinogens, which could be the commercial end of CNTs [24]. However, especially health risks of CNTs are widely discussed in the community. One key point for further understanding could be the surface termination of CNTs. It should be underlined that there are only indicators available that indicate the possible risk of this material. Furthermore, it should be noted that CNTs will most likely have no influence on health if they are incorporated as fillers in composite materials like polymers as long as there is no abrasion. Up to now, CNTs and health are still an open question that is widely discussed. However, concrete results are still missing.

Nanodiamond thin films are supposed to be completely safe [5] without any negative impact on health. Nanodiamond scalpels are used for eye surgery applications in medicine. However, the production process of diamond is relatively slow and requires a significant amount of energy, which makes these films relatively expensive. Furthermore, CVD of diamond was a scientific hype in the 1990s. After missing success stories and products many companies left this field. Some of these companies concentrated on diamond-like carbon films, which are currently more mature, at least for hard coatings. Therefore, the topic of diamond is sometimes considered an exotic field, which makes it more difficult to realize real-world applications. Therefore, the "new material" nanodiamond was clearly overlooked. The physical and chemical properties of this material are clearly outstanding and extreme and offer therefore a wide range of possible applications in the future.

8.3.3 Boundary Conditions

An extremely high modulus and elastic strain, in addition to a tensile strength in an order of magnitude higher than that of conventional carbon fibers, make CNTs the ultimate reinforcement in polymer matrix composite materials and perhaps also in light metals. Additionally, CNTs may be capable of meeting requirements for antielectrostatic fillers for an insulating polymer matrix or for applying electrostatic painting processes. The use of CNTs as filler materials is expected to increase the electrical conductivity of polymers, together with the desired improvement in mechanical properties. Given the predicted two-digit annual average growth rate of polymer nanocomposites, this seems to be a very attractive field in the short term. However, the success of CNTs in this area will also depend on their price competitiveness compared to other filler materials such as layered silicates, metal nanoparticles, or nanofibers.

Other possibly attractive markets for CNTs include fillers for paints and lacquers in order to increase electrical conductivity and to avoid electrostatic charging. As already mentioned above, composites filled with CNTs can be used for lightweight tanks and pipes (e.g., for automotive applications). In the case of composites advantage can be taken from the electrical, mechanical, and thermal properties of CNTs. The advantage of taking CNTs as filler materials is that only small amounts of CNTs have to be used. This fact can compensate the high costs of CNTs (cf. Fig. 8.2).

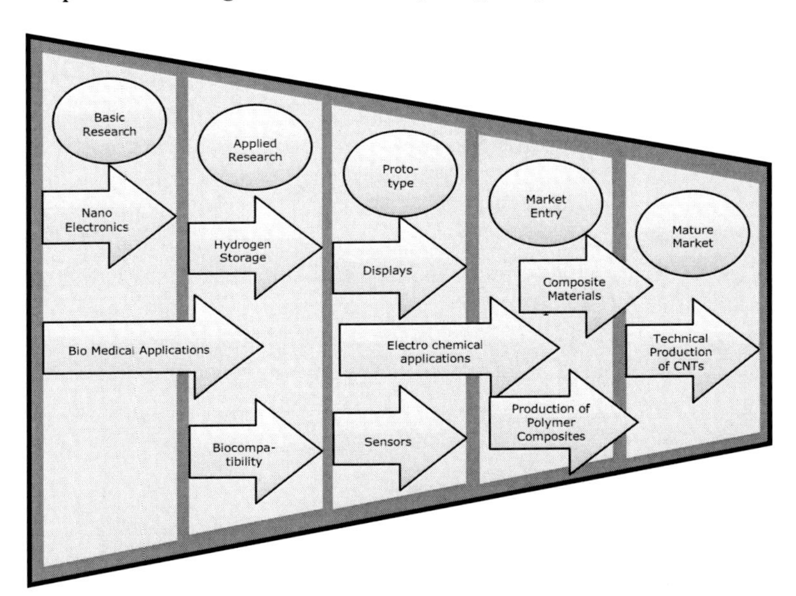

Figure 8.2 Value-added chain of CNTs and nanodiamond.

Nanofibers are generating great interest in certain industry segments, where alternative materials are characterized by limited performance or much higher unit prices.

8.3.4 Economic Information and Analysis

Carbon-based materials offer a wide range of possible applications. However, carbon as a material with a number of potential applications has been frequently overlooked in the EU. Japan concentrates research and funding on the application of several carbon-based materials. A number of Japanese funding approaches illustrate that; one example is funding of CNT research with about €15 million per year [25]. However, even in Europe CNT research has been intensified during recent years. This is indicated among other things by large national funding and innovation programs such as "Genesis" in France, which started in 2008, or "Innovation Alliance CNT" initiated by the German Ministry of Education and Research in 2008/2009.

The current market for CNTs is estimated to total approximately US$10 million for predominantly research-grade nanotubes with production quantities of several kilograms [26]. The partly very optimistic future prospects of nanotubes have to be treated very carefully. In most applications discussed, the future progress of CNTs in mass markets depends critically on their price development and price competitiveness compared with alternative materials. Global players like Mitsui & Co., Japan, have announced the start of mass production in the range of 120 tons per year. This massive increase in CNT supply will certainly contribute to bringing down prices and promote the substitution of conventional materials like carbon fibers by CNTs. In a similar way Bayer MaterialScience (Germany) very recently started a CNT production plant with a capacity of 200 tons per year and wants to increase this to 3,000 tons per year in the medium term. Arkema (France) announced the construction of a CNT pilot production plant based on entirely biosourced raw material with a capacity of 400 tons per year starting from 2011.

The application and possible end markets of CNTs and nanodiamond in various fields are shown in Figs. 8.2 and 8.3. It can be seen as a value-added chain where the products using CNTs and nanodiamond are listed according to their stadium of application. CNTs, for example, can be used for sport goods, transistors, and flat screens.

8.3.5 Patent Analysis

8.3.5.1 Carbon nanotubes

To get an overview of patents with reference to CNTs, which were published during the last few years, both the patent database of the

World Intellectual Property Organization (WIPO) and the European Patent Office's (EPO) free online service "esp@cenet" were analyzed. "esp@cenet" showed 3,500 hits regarding European patent applications [27]. Figures 8.4 and 8.5 focus on WIPO applications.

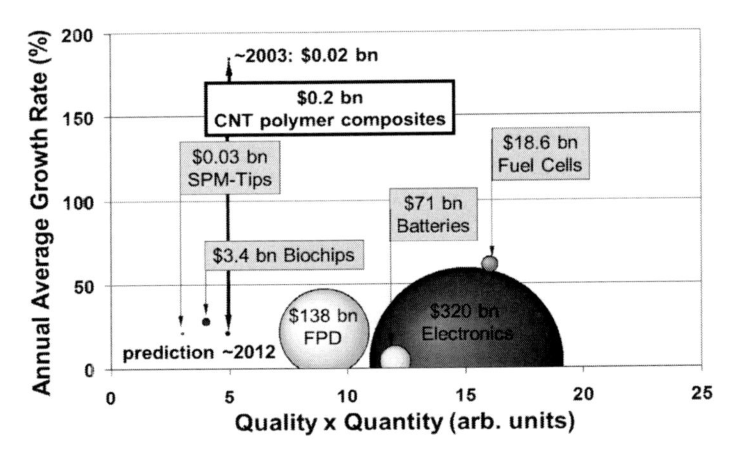

Figure 8.3 Market sizes and growth rates versus market requirements of possible end markets for CNT products. It is always the whole world market size given. Most of the markets were up to now not captured by CNTs (except the world market for CNT–polymer composites.)

As shown in Fig. 8.4, the number of patent applications published each year was nearly constantly growing from 1 in 1995 to 309 in 2008. The number of applications in 2009 was not fully available during the time of the patent analysis.

The WIPO patentscope [28] offers the possibility of a country assignment of patent applications to the first applicant's country, giving a rough impression on where the new invention has been made. Figure 8.5 illustrates how many patents were announced by the listed countries from 1995 until now.

According to Fig. 8.5, the United States published about half of the patents during the last 14 years, in comparison to the rest of the world. It was followed by Japan, South Korea, France, and Germany. If the numbers of inventions made by all of the European countries (France, Germany, Great Britain, the Netherlands, and Belgium) are summarized into one representative amount, it can be seen that Europe takes third place after the United States and Japan.

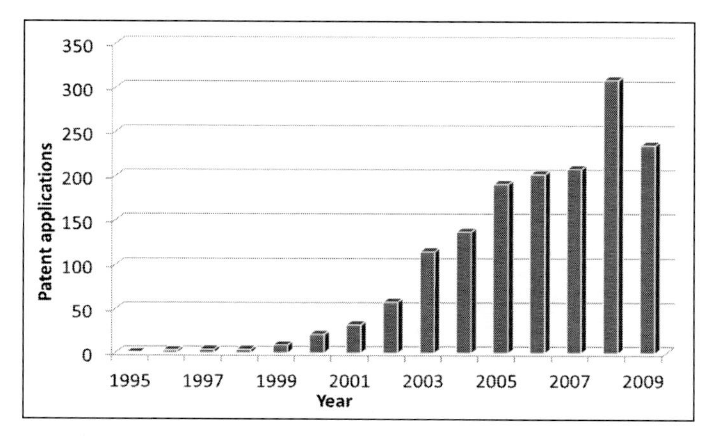

Figure 8.4 Number of patent applications per year concerning CNTs (*Source:* WIPO patenscope; keyword search; date of data acquisition: November 29, 2009).

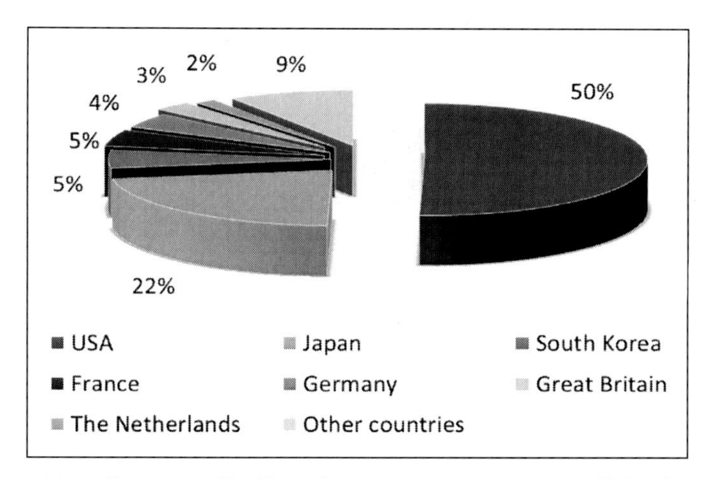

Figure 8.5 Patent applications by countries concerning CNTs (*Source:* WIPO patenscope; keyword search; date of data acquisition: November 29, 2009).

8.3.5.2 Nanodiamond

Nanodiamond or ultrananocrystalline diamond (UNCD) is currently less attractive when compared to CNTs, at least according to the number of patent applications. The database of the European Patent Office shows only 113 hits when searched for patent applications with variations of the keyword "nanodiamond."

An overview of international patent applications in the field of nanodiamond concerning their quantity during the last few years, as well as their country of origin, is visualized in Figs. 8.6 and 8.7. The relating figures have been extracted from the patent database of WIPO.

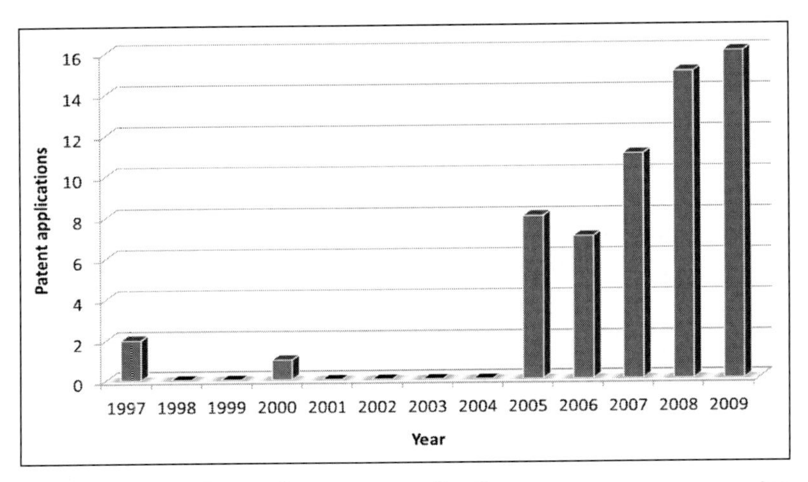

Figure 8.6 Number of patent applications per year concerning nanodiamond (*Source:* WIPO patenscope; date of data acquisition: November 21, 2009).

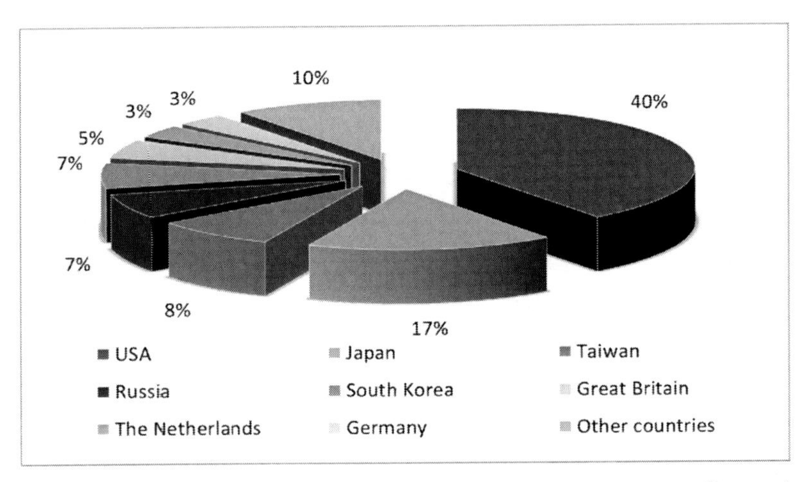

Figure 8.7 Patent applications by countries concerning nanodiamond (*Source:* WIPO patenscope; keyword for the advanced search: nanodiamond; date of data acquisition: November 21, 2009).

Looking at the number of published patents per year it can be asserted that the amount of patents was nearly not recognizable until 2005. Afterward there was a more or less constant growth of the number of patents until now.

Figure 8.7 illustrates that the United States published most of the patents since 1993, followed by Japan and Taiwan. Summarizing the number of patents in Europe during the last 16 years results in the second position for Europe, directly following the Unites States and ranking in front of Japan.

Acknowledgments

We would like to thank Elenour O'Rouke (Institute of Nanotechnology, U.K.) for helpful discussions. Further, we would like to thank Ingo Barbré from the NMTC, Berlin, Germany, and Peter Gluche from GFD mbH, Ulm, Germany. The financial support within the European project "ObservatoryNANO" from the European Commission is highly acknowledged. More information can be found on the project website (http://www.observatory-nano.eu/project/.) The passages are partly published on the "ObservatoryNANO" web page.

References

1. Brand, L., Gierlings, M., Hoffknecht, A., Wagner V., and Zweck, A. (2009). *Kohlenstoff-Nanoröhren—Potenziale einer neuen Materialklasse für Deutschland*, Schriftenreihe Zukünftige Technologien Band 79, ISSN: 1436-5928 (VDI-Technologiezentrum Düsseldorf, Germany).
2. Baughman, R.H., Zakhidov, A., and de Heer, W.A. (2002). Carbon nanotubes—the route toward applications, *Science*, **297**, pp. 787–792.
3. Coleman, J.N., Khan, U., and Gun'ko, Y.K. (2006). Mechanical reinforcement of polymers using carbon nanotubes, *Adv. Mater.*, **18**, pp. 689–706.
4. http://www.electronics.ca/presscenter/articles/743/1/Carbon-Nanotube-Production-Dramatic-Price-Decrease-Down-to-150 kg-for-Semi-Industrial-Applications/Page1.html (Feb. 6, 2008).
5. Interview result.
6. Wagenknecht, U., Kretzschmar, B., Pötschke, P., Costa, F.R., Pegel, S., Stöckelhuber, K.W., and Heinrich, G. (2008). Polymere Nanokomposite mit anorganischen Funktionsfüllstoffen, *Chem. Ing. Tech.*, **80**, pp. 1683–1699.

7. "Plastics processing technologies 2008," VDI conference, Cologne, Germany.
8. Thayer, A.M. (2007). Carbon nanotubes by the metric ton, *Chem. Eng.*, **85**, pp. 29–35.
9. Cheaptubes. (2009). http://www.cheaptubesinc.com/.
10. Arry Nanomaterials. (2009). http://www.arry-nano.com/.
11. Electronics.ca. (2008). Carbon Nanotube Production Dramatic Price Decrease Down to 150kg for Semi Industrial Applications.
12. Maurer, J. (2008). Starkes Interesse an der Nanotechnologie, *Asien Kurier*, **3**, pp.18–20.
13. Chemie.de Information Service. (2009). Arkema to Build Carbon Nanotube Pilot Plant in France.
14. http://www.future-carbon.de/index.php?id=45.
15. http://www.empa.ch/plugin/template/empa/1173/*/---/l=1.
16. Interview result.
17. The carbon black price depends very much on the quality. Therefore, the price can be only considered as a guideline.
18. The gold price showed a dramatically increase over the past few months due to the economic crisis. The value was taken from Bundesverband Deutscher Banken. The gold value is US$ 870 per ounce (compared to US$ 272 per ounce in the year 2000.)
19. http://www.bccresearch.com/report/NAN024C.html (Nov. 6, 2009).
20. http://www.spiegel.de/wissenschaft/mensch/0,1518,656362,00.html (Nov. 6, 2009).
21. Salonen, E., Lin, S., Reid, M.L., Allegood, M., Wang, X., Rao, A.M., Vattulainen, I., and Ke, P.-C. (2008). Real-time translocation of fullerene reveals cell contraction, *Small*, **4**, pp. 1986–1992.
22. Harper, T. (2009). Nanotechnologies in 2009—Creative Destruction or Credit Crunch? (Cientifica, Ilkley, U.K.), http://www.cientifica.eu/attachments/054_Nanotechnologies%20in%202009.pdf (Nov. 6, 2009).
23. http://www.ivam.de/index.php?content=mitteilung_details&mitteilung_id=1424&typ=presse (Nov. 6, 2009).
24. http://epa.gov/ncer/publications/workshop/10_26_05/abstracts/hurt.html (Nov. 6, 2009).
25. Brand, L., Gierlings, M., Hoffknecht, A., Wagner, V., and Knecht, A. (2009). *Kohlenstoff-Nanoröhren: Potenziale einer neuen Materialklasse*

für Deutschland (VDI Technologiezentrum, Düsseldorf, Germany) (in German).

26. Interview result.
27. http://ep.espacenet.com/ (Nov. 16, 2009).
28. WIPO patentscope, http://www.wipo.int/pctdb/en/.

Index